ARTIFICIAL INTELLIGENCE FOR FUTURE INTELLIGENT TRANSPORTATION

Smarter and Greener Infrastructure Design

ARTIFICIAL INTELLIGENCE FOR FUTURE INTELLIGENT TRANSPORTATION

Smarter and Greener Infrastructure Design

Edited by

Rajesh Kumar Dhanaraj, PhD

Nilayam Kumar Kamila, PhD

Subhendu Kumar Pani, PhD

Balamurugan Balusamy, PhD

Vani Rajasekar, PhD

AAP | APPLE ACADEMIC PRESS

First edition published 2024

Apple Academic Press Inc.
1265 Goldenrod Circle, NE,
Palm Bay, FL 32905 USA

760 Laurentian Drive, Unit 19,
Burlington, ON L7N 0A4, CANADA

CRC Press
2385 NW Executive Center Drive,
Suite 320, Boca Raton FL 33431

4 Park Square, Milton Park,
Abingdon, Oxon, OX14 4RN UK

Library and Archives Canada Cataloguing in Publication

Title: Artificial intelligence for future intelligent transportation : smarter and greener infrastructure design / edited by Rajesh Kumar Dhanaraj, PhD, Nilayam Kumar Kamila, PhD, Subhendu Kumar Pani, PhD, Balamurugan Balusamy, PhD, Vani Rajasekar, PhD.
Names: Dhanaraj, Rajesh Kumar, editor. | Kamila, Nilayam Kumar, editor. | Pani, Subhendu Kumar, 1980- editor. | Balusamy, Balamurugan, editor. | Rajasekar, Vani, editor.
Description: First edition. | Includes bibliographical references and index.
Identifiers: Canadiana (print) 20230495818 | Canadiana (ebook) 20230495850 | ISBN 9781774913529 (hardcover) | ISBN 9781774913536 (softcover) | ISBN 9781003408468 (ebook)
Subjects: LCSH: Intelligent transportation systems. | LCSH: Artificial intelligence.
Classification: LCC TE228.3 .A78 2024 | DDC 388.3/12—dc23

Library of Congress Cataloging-in-Publication Data

CIP data on file with US Library of Congress

ISBN: 978-1-77491-352-9 (hbk)
ISBN: 978-1-77491-353-6 (pbk)
ISBN: 978-1-00340-846-8 (ebk)

About the Editors

Rajesh Kumar Dhanaraj, PhD
Associate Professor, School of Computing Science and Engineering, Galgotias University, Greater Noida, U.P, India

Rajesh Kumar Dhanaraj, PhD, is an Associate Professor in the School of Computing Science and Engineering at Galgotias University, Greater Noida, U.P., India. He has published over 35 articles and papers in various journals and conference proceedings and contributed chapters to books. His research and publication interests include cyber-physical systems, wireless sensor networks, and cloud computing. He is an expert advisory panel member of Texas Instruments Inc., USA. He holds a PhD degree in Information and Communication Engineering from Anna University Chennai, India.

Nilayam Kumar Kamila, PhD
Manager, Platform & Data Architecture Engineering Stream, Retail & Direct Technology, Capital One, USA

Nilayam Kumar Kamila, PhD, is an experienced Technical Architect and Lead, currently working as Manager in Platform and Data Architecture Engineering Stream under Retail and Direct Technology in Capital One, USA. He is currently an active member of IEEE and imparting his profound knowledge in the fields of science and engineering. He received prestigious professional certifications in machine learning, neural networks, and deep learning from Stanford University, USA. He completed his MS in Science, Master in Computer Application, and Master of Technology in Computer Science from Utkal University, India. He was awarded his PhD from Sri University, India, in 2017, in Computer and Information Systems.

Subhendu Kumar Pani, PhD
Professor, Krupajal Engineering College, Odisha, India

Subhendu Kumar Pani. PhD, received his PhD from Utkal University, Odisha, India, in the year 2013. He is working as a Professor at Krupajal Engineering College under BPUT, Odisha, India. He has more than 18 years of teaching and research experience. His research interests include data mining, big data analysis, web data analytics, fuzzy decision making, and computational intelligence. He is the recipient of five researcher awards. In addition to research, he has guided two PhD students and 31 MTech students. He has published 51 international journal papers (25 Scopus-indexed). His professional activities include roles as book series editor (CRC Press, Apple Academic Press, Wiley-Scrivener), associate editor, editorial board member, and/or reviewer of various international journals.

Balamurugan Balusamy, PhD
Professor, School of Computing Sciences and Engineering at Galgotias University, Greater Noida, India

Balamurugan Balusamy, PhD, is currently working as a Professor in the School of Computing Sciences and Engineering at Galgotias University, Greater Noida, India. His contributions focus on engineering education, blockchain, and data sciences. His academic degrees and 12 years of experience working as a faculty in a global university like VIT University, Vellore, has made him more receptive and prominent in his domain. He has 200 plus high-impact factor papers in Springer, Elsevier, and IEEE. He has done more than 50 edited and authored books and collaborated with eminent professors across the world from top QS-ranked universities. Prof. Balamurugan Balusamy has served up to the position of Associate Professor in his stint of 12 years of experience with VIT University, Vellore.

Vani Rajasekar, PhD
Assistant Professor, Department of Computer Science and Engineering, Kongu Engineering College, Tamil Nadu, India

Vani Rajasekar, PhD, completed BTech (Information Technology), and MTech (Information and Cyberwarfare) in the Department of Information and Technology, Kongu Engineering College. She completed her PhD (Information and Communication Engineering) in the area of Biometrics and Network Security. Presently, she is working as an Assistant Professor in the Department of Computer Science and Engineering at Kongu Engineering College. Her area of interest includes cryptography, biometrics, network security, and wireless networks. She has authored around 25 research papers and book chapters published in various international journals and conferences, which were indexed in SCI, Scopus, and Web of Science.

Contents

Contributors

Yusuf Jibrin Alkali
ICT Support Officer, Federal Inland Revenue Service, Nigeria

S. Anbukkarasi
Assistant Professor, Department of Computer Science and Engineering, Kongu Engineering College, Tamil Nadu, India

Dev Aroa
School of Computing Science and Engineering, Galgotias University, Greater Noida, Uttar Pradesh, India

Devansh Arora
School of Computing Science and Engineering, Galgotias University, Greater Noida, Uttar Pradesh, India

K. S. Arvind
Associate Professor, Department of Computer Science and Engineering, JAIN University, Bangalore, Karnataka, India

Rajesh Kumar Dhanaraj
Associate Professor, Computer Science and Engineering, School of Computing Science and Engineering, Galgotias University, Uttar Pradesh, India

C. R. Dhivyaa
Assistant Professor, Department of Computer Science and Engineering, Kongu Engineering College, Tamil Nadu, India

Gaurav Dhuuriya
School of Computing Science and Engineering, Galgotias University, Greater Noida, Uttar Pradesh, India

Shashivendra Dulawat
Civil Engineering Department, Mewar University, Chittorgarh, Rajasthan, India

N. R. Gayathiri
Department of Artificial Intelligence and Data Science, Bannari Amman Institute of Technology, Erode, Tamil Nadu, India

T. Janani
Assistant Professor, Bannari Amman Institute of Technology, Sathyamangalam, Tamil Nadu, India

Prashant Johri
School of Computing Science and Engineering, Galgotias University, Greater Noida, Uttar Pradesh, India

M. Kalamani
Department of Electronics and Communication Engineering, KPR Institute of Engineering and Technology, Tamil Nadu, India

A. Kodieswari
Department of Information and Technology, Bannari Amman Institute of Technology, Erode, Tamil Nadu, India

M. Krishnamoorthi
Department of Information Technology, Dr. N.G.P. Institute of Technology, Tamil Nadu, India

K. Sathis Kumar
Assistant Professor, Bannari Amman Institute of Technology, Sathyamangalam, Tamil Nadu, India

Yasemin Sarikaya Levent
Akdeniz University, Antalya, Turkey; City and Regional Planning Department, Mersin University, Turkey

Salim Idris Malami
School of Energy, Geo-Science, Infrastructure and Society, Institute for Sustainable Built Environment, Heriot-Watt University, Edinburgh, United Kingdom; Civil Engineering Department, Kano University of Science and Technology, Wudil, Kano, Nigeria

T. Malathi
Department of Artificial Intelligence and Data Science, Bannari Amman Institute of Technology, Erode, Tamil Nadu, India

S. Malathy
Assistant Professor, Department of Computer Science and Engineering, Kongu Engineering College, Erode

Auwal Alhassan Musa
Civil Engineering Department, Mewar University, Chittorgarh, Rajasthan, India;
Civil Engineering Department, Kano University of Science and Technology, Wudil, Kano, Nigeria

S. Mythili
Department of Electronics and Communication Engineering, Bannari Amman Institute of Technology, Tamil Nadu, India

Rahul Reddy Nadikattu
Senior IEEE Member, Department of IT, University of Cumbersome, USA

S. Naveena
Assistant Professor, Bannari Amman Institute of Technology, Sathyamangalam, Tamil Nadu, India

K. Nithya
Department of Computer Science and Engineering, Kongu Engineering College, Tamil Nadu, India

S. Nithyapriya
Department of Artificial Intelligence and Data Science, Bannari Amman Institute of Technology, Erode, Tamil Nadu, India

P. Ponnila
Assistant Professor, Bannari Amman Institute of Technology, Sathyamangalam, Tamil Nadu, India

T. C. Kalaiselvi
Professor, Department of Electronics and Communication Engineering, Kongu Engineering College, Perundurai, Tamil Nadu, India

K. M. Karthick Raghunath
Associate Professor, Department of Computer Science and Engineering, Jain University, Bangalore, Karnataka, India

Tanul Prabhat Rai
School of Computing Science and Engineering, Galgotias University, Greater Noida, Uttar Pradesh, India

K. Remya
Department of Electronics and Instrumentation Engineering, Bannari Amman Institute of Technology, Erode, Tamil Nadu, India

K. R. Sabarmathi
Department of Information and Technology, Bannari Amman Institute of Technology, Erode, Tamil Nadu, India

Kabeer Tijjani Saleh
Akdeniz University, Antalya, Turkey

C. Sharmia
Assistant Professor, Department of Computer Science and Engineering, Kongu Engineering College, Perundurai, Tamil Nadu, India

Shubhang Sinha
School of Computing Science and Engineering, Galgotias University, Greater Noida, Uttar Pradesh, India

P. Sobiyaa
Assistant Professor, Bannari Amman Institute of Technology, Sathyamangalam, Tamil Nadu, India

K. S. Suganya
Assistant Professor, Department of Information and Technology, Bannari Amman Institute of Technology, Sathyamangalam, Tamil Nadu, India

C. N. Vanitha
Assistant Professor (Sr. G), Department of Computer Science and Engineering, Kongu Engineering College, Erode, Tamil Nadu, India

Arun Velu
Director Equifax and Researcher, Atlanta, USA

R. Vinothsaravanan
Assistant Professor, Department of Computer Science and Engineering, JAIN University, Bangalore, Karnataka, India

Pawan Whig
Senior IEEE Member, Dean Research, Vivekananda Institute of Professional Studies, New Delhi, India

Abbreviations

ACO	ant colony optimization
AI	artificial intelligence
AIS	artificial immune system
ANN	artificial neural networks
AQI	air quality index
AQM	air quality monitoring
AQS	air quality system
ATIS	advanced traveler information system
ATMS	advanced traffic management system
AV	autonomous vehicles
BCO	bee colony optimization
Bi-LSTM	bi-directional long short-term memory
BMS	bridge management system
B-RBES	belief rule-based expert system
BRT	bus rapid transit
CCRAQM	central control room for air quality management
CNN	convolutional neural network
CO	carbon monoxide
DP	dynamic programming
EEG	electroencephalogram
EIA	environmental impact assessments
EL	ensemble learning
ELRT	elevated light rail transit
EMMPF	efficient multiple model particle filter
EOG	electrooculogram
ERP	enterprise resource planning
ERTMS	European Rail Traffic Management System
ES	expert system
ETA	estimated times of arrival
EVs	electric vehicles
GA	genetic algorithm
GFEI	global fuel economy initiatives
GHGs	greenhouse gas

GNN	graph neural network
GPS	global positioning system
GRU	gated recurrent unit
GUI	graphical user interface
HCs	hydrocarbons
IC	integrated circuit
IEC	International Electrotechnical Commission
IMO	International Maritime Organization
IoT	internet of things
IR	infrared sensor
ITS	intelligent transportation systems
JPO	joint program office
KB	knowledge base
LiDAR	light detection and ranging
LRT	light rail transit
LSTM	long short-term memory
MAE	mean absolute error
MAPE	mean absolute percentage error
MB-SGD	mini-batch stochastic gradient descent
MCMS	micro-climate monitoring system
ML	machine learning
MPE	mean prediction error
NAAQS	National Ambient Air Quality Standards
NAG	Nesterov Accelerated Gradient
NDP	network design problem
NLL	negative-log likely
NLP	natural language processing
NMT	non-motorized transport
NO_2	nitrogen dioxide
NOx	nitrogen oxides
O_3	ozone
PAHs	polycyclic aromatic hydrocarbons
PBS	pattern-based strategy
PCA	principal component analysis
PM	particle matter
PMS	pavement management system
PRC	People's Republic of China
PTC	positive train control

RADAR	radio detection and ranging
RFID	radio frequency identification
RL	reinforcement learning
RMSE	root mean squared error
RNN	recurrent neural network
RoI	region of interest
RP	recurrence plot
RPE	relative prediction error
RPM	revs per minute
SA	simulated annealing
SEPTA	Southern Pennsylvania Transportation Authority
SGD	stochastic gradient descent
SMART	Suburban Mobility Authority for Rapid Transit Authority
SMRTL	Singapore Mass Rapid Transit lines
SMS	short message service
SNN	stimulated neural networks
SO_2	sulfur dioxide
SOA	secondary organic aerosol
SO_x	sulfur oxide
STS	smart transportation system
SVM	support vector machine
TIMS	transportation infrastructure management system
TIS	transportation infrastructure systems
TOD	transit-oriented development
UAVs	unmanned aerial vehicles
UI	user interface
V2I	vehicle-to-infrastructure
VCS	vehicle control system
VDS	vehicle driving simulator
VES	vehicle emission standards
VO	visual odometry
VOCs	volatile organic compounds
VPN	virtual private network
WHO	World Health Organization
WLAN	wireless local area network

Preface

Artificial intelligence (AI) is advancing at a breakneck pace, presenting unprecedented prospects to improve the performance of various industries and enterprises, including the transportation sector. Highly advanced computing approaches that imitate the way the human brain operates are among the AI advancements. The use of artificial intelligence in transportation aims to address the difficulties of rising travel demand, CO_2 emissions, safety concerns, and environmental degradation. In this digital age, resolving these challenges in a more efficient and effective manner has become more conceivable due to the availability of a vast amount of quantitative and qualitative data, as well as AI.

Artificial neural networks (ANN), genetic algorithms (GAs), simulated annealing (SA), artificial immune system (AIS), ant colony optimizer (ACO), bee colony optimization (BCO), and fuzzy logic model are examples of AI approaches that are finding their way into the transportation area (FLM). A good grasp of the links between AI and data, on the one hand, and transportation system characteristics and factors, on the other, is required for successful AI implementation. Furthermore, it is hoped for transportation authorities to establish how to employ these technologies to quickly improve congestion relief, make travel time more predictable for their consumers, and improve the economics and efficiency of their critical assets.

Transportation experts and professionals are facing more complicated issues at the turn of the 21st century. They are challenged to provide safe, efficient, and dependable transportation while reducing the impact on the environment and communities. In recent years, transportation researchers and practitioners have become more interested in examining the feasibility of using AI paradigms to address some of the aforementioned issues in order to improve transportation systems' efficiency, safety, and environmental compatibility. In the 1950s and 1960s, AI researchers frequently set grandiose goals for the subject, such as the development of general-purpose problem solvers. Our goal in investigating AI applications for transportation as transportation researchers and professionals, on the other hand, is far more modest.

The book presents a comprehensive review of the difficulties and recent developments in the subject of smart mobility and transportation encompassing technical, algorithmic, and social elements. It examines innovative service and platform concepts for future mobility. Artificial intelligence principles are examined, as well as their implementation in modern hardware architecture. New machine learning (ML) issues for autonomous vehicles (AVs) and fleets are investigated in the framework of artificial intelligence. In addition, the book looks into the human dynamics and social implications of future mobility concepts. The purpose of this book is to present a comprehensive approach to smart transportation.

This book presents an overview of AI strategies used to solve transportation challenges around the world, with a focus on traffic management, traffic safety, public transportation, and urban mobility. Finally, the overview discusses the challenges and limitations of AI applications in transportation. The book examines modern technologies such as cloud computing, machine learning, and communication in the context of fully atomized transportation that meets residents' requirements. This will result in a complete shift in transportation concepts. Highlighting most of the research directions in today's security era, this book is more suitable for researchers from cyber security, machine learning analysis, data analysts, AI researchers, ethical hackers, students, and academicians.

CHAPTER 1

Introduction to Artificial Intelligence in Smart Transportation

GAURAV DHUURIYA, SHUBHANG SINHA, DEV AROA, TANUL PRABHAT RAI, PRASHANT JOHRI, and DEVANSH ARORA

School of Computing Science and Engineering, Galgotias University, Greater Noida, Uttar Pradesh, India

ABSTRACT

Artificial intelligence (AI) has already impacted the ways of the world in which we live, and its impact is not slowing down but rather increasing at a drastic speed. It has created a world where the set of technologies are helping humans to solve problems very easily. AI has a major number of applications in several fields. One such field is transportation, where it has already created a major impact on how we move people and goods from one point to another. AI has helped us in transportation in far many ways from decreasing the number of accidents to traffic monitoring. AI has helped us to provide more safer, reliable, and efficient transportation. Not only that, AI has enabled us to provide more economical and sustainable transportation which not only has resulted in economic boom but also it has resulted with much cleaner environment. Its learning behavior has led to improvement in computation, monitoring but also in communication.

Its way of handling things and learning of new approaches has been exceptional and lead to many advances. Since its introduction, these last few decades have seen a boom in its growth and its aim to improve has been tenfold. AI leads to innovation and innovation leads to a better way

Artificial Intelligence for Future Intelligent Transportation: Smarter and Greener Infrastructure Design. Rajesh Kumar Dhanaraj, Nilayam Kumar Kamila, Subhendu Kumar Pani, Balamurugan Balusamy, & Vani Rajasekar (Eds.)

of doing something, same can be said for transportation, whether it is tesla self-driving cars or AI-driven robots which may act as traffic enforcers in future and reduce the cost of labor, in this chapter we will look into details and some already existing applications of AI in the world of transport with respect to various changing factors and parameters. AI has strong set of subsets which can be used to further elaborate its power, one of these subsets is machine learning (ML) which is the art of programming a system by feeding it data and letting it work on its own. This can be implemented into the world of transportation as well. In layman's term, the possibilities with AI are endless and we must use it to our advantage.

1.1 INTRODUCTION

Every day, the world we live-in is changing at a rapid pace, irrespective of the area or sector, there are heavy innovations which are making our work easier. AI is playing a huge part in this ever-changing world, whether it is sports, medical areas, or anything else AI is at the heart of its change. Transport is no different than these sectors, AI is capable of making it better than ever.

Artificial intelligence (AI) in transpiration has helped us to compete in many tasks such as self-driving car, directing person to his/her destination, provide most suitable path for the journey, it has led to reduction to less accident, less wastage of time, has prevented my accident, provided us forecast of weather, traffic, and majorly led to prevention of human error. AI has helped in damages caused due to human error which has us to increase at exponential rate. Now we don't need to deal with unpredictability as almost every prediction can be made now with the help of AI.

AI has majorly impacted the security around the globe, which has resulted in much safer cities and locality. Transportation security has increased my folds and tracking of crime is much easier, which has boosted public opinion about public transports.

AI is already getting used in places with fixed-track transportation, such as railways and metros. AI is providing its assistance in this type of transportation in my ways, such as providing availability of seats, routes, and other features on the internet and even at devices inform of app and all. AI can also help in providing proper guidance to passenger about their destination and on notifying them on reaching there stop or station. It can lead to advancement where a train or metro is not getting man-handled and

AI is driving these transports which would be more convenient as there will be less amount of human error. In future, we could loss imagine that these whole transport department is being handled by AI from providing ways to transporting people.

Air and Sea Transpiration has also been benefited by AI but not as much as Road Transpiration but still it has led to a major impact in this field. Such is the case of air transpiration where AI has helped in security as tracking of criminal from fleeing somewhere with the help of facial Recognition and other application of AI. It has also helped in tracking of illegal smuggling also. These are only few benefits mentioned but AI has led to uncountable benefits in these field of transport and it has a bright future in this field such as in case of water transport AI could help us in managing the navigation part which includes whole route travel and even help in transmitting of location without any human help and even maybe anchor the ship on its own. AI is craving at a path in air transport where whole flight will be piloted by ai alone or by single pilot as AI could pilot the plan at auto pilot now also.

Road transportation has been a place where AI has most strong hold and been in work since past two decades where the work on providing assistance in uncountable number of matters comprising from providing routes to monitoring traffic. The impact of AI on road transportation has been large, and it has changed the way people look towards the AI. There is various application of AI in this filed such as most basic one where they provide there assistance to driver where they guide them to their destination saving their time and effort as they provide the most reliable closest most efficient path for them preventing them to go through an unnecessary traffics. AI also helps authority to monitor vehicles so if they violate any rules, they could be dealt with. One of the most promising applications in this filed is driverless cars which had been highlights in this decade and they are most promising thing of AI in transportation, being able to travel without driving has been a dream, which is going to come to present. This could help in not only personal transportation but with public transport as well. Driverless vehicle is on its way to us and will be available for us in a short time.

There are many ways in which AI could be useful in field of transportation such as providing assistance and medical emergency to anyone who had been in accident as AI to alert authority in case of emergency. AI could also help in guiding people at a very easy pace so that they could cope up with it and thus decreasing waste of any human hours. It could even help in repair which will lead to prevention of losing any human life (Figure 1.1).

FIGURE 1.1 Artificial intelligence in smart transportation.

With AI in working transportation is more convenient, more reliable, more active, which has led to human to live more time to spare which was saved by them in transport. AI is a key to next generation of evolution in transport which could lead to more agile and intelligent transportation in the future.

1.2 AI IN TRANSPORTATION

1.2.1 GLOBAL MARKET

AI has drastically affected the world's economy and it is continuing to do so at the much faster pace. AI is going at a much faster pace and will be able to get hold of $1 trillion of the global revenue by the midst of the next decade. Last year, AI has increased its revenue from more a billion dollar to three dollars in the last two years alone while, it is predicted it will continue to grow like this and will reach a market share of $5 million by the end of 2024. The AI in transportation has been exponential and with its increased production it will be increase my fold in upcoming year. USA being the most influential among them all and increased sales by more than half of 100%.

1.2.2 OPPORTUNITIES CREATED THROUGH AI INTERVENTIONS

AI has uncountable number of applications in field of transportation one of them being autonomous vehicles (AVs) which has most influence in this field. Though other application is much more applicable in society, and they have a huge impact there, they are solving day-to-day struggle faced by human and they are overcoming them for the sake of humans and helping transportation to go much more easily and conveniently.

Through AI help, it is much easier to predict something more conveniently and it has become a more reliable source to work with. With help of application and internet AI can provide us best routes, paths to travel and cover our distance inform of Google Maps, Apple Map and many more. Thus, it has increased much satisfaction level among the users.

With help of AI-based application such as Uber, Ola, and Rapido a mode of transport can be at your home to take you to your destination just by one click. Which is more secure, economical and can help increase public transport globally [1]. There are many places where it is getting used one being USA's economical hub NY where they have planned to lay smart bus track and make public transport more convenient and reliable for the people [15].

1.2.3 EFFICIENCY

One of the most applicable applications of AI is security, road traffic deaths lead to almost 2 million deaths in a year. This was half in year 2011 and it doubled in a decade. Thus, we need application of AI to work to save people from these number of deaths which happens to poor infrastructure and human error and can lead to unnecessary deaths in the meantime due to not proper conveying of data regarding accidents but with help of AI we are overcoming these problems as AI application convey message to nearest authority is something unusual happens. Lastly self-driving car are being developed by tested and it will take over world with a storm and it is said that 1 in every 4 cars will be Autonomous Vehicle by the midst of next decade with a decrease in emission of gases globally [16].

AI can reduce the amount of emission at a very high scale by providing most suitable paths or routes, which led to fuel efficiency hence can led to less emission of toxic gases therefore preventing global warming [2].

There are eco-friendly AI trucks which are not only good for traffic management but are also helping our environment to stay green and healthy.

1.3 ARTIFICIAL INTELLIGENCE (AI) IN ROAD TRANSPORTATION

AI in road transportation is one field where AI is used at maximum level from guiding ways to monitoring of vehicles. There are an uncountable number of applications of AI in a single vehicle that are being used to help human to ease a little bit. AI in road transportation is at booming phase and everyone to explore this field, which has resulted in great economic generation. Automotive, technical, and investigations firms are trying to develop applications, sensors, command unit and software that make use of AI in that field. These has led to plenty amount of application of transportation in market increasing revenue and easing life of people [6].

There are many applications of AI in road transport one being automatic vehicles which are being used for commercial purposes. Not like autonomous motor vehicles these vehicles working in clutters where they compromise if group for trucks and named as truck platooning. In this, there are groups for trucks which are getting handled by AI and are maintain some distance for safe transport of goods. This is getting tested globally for increment of amount of transferrable goods at one time. These truck platooning could be handled by one single driver with the help of AI. In this, driver does not have to drive all the time but driving could be handled by AI alone and at sometimes driver need to indulge with it rest of the journey driver has just to monitor things for safekeeping (Figure 1.2).

There are also may benefits of AI in transportation such as booking of rides such as Ola, Uber, Rapido which has helped user not only to worry about boarding transport from somewhere rather than it has helped to providing services at their doors and wherever they want [15]. People now can share rides and even book their own personal rides with n number of possibilities of the type of vehicles they want. This has helped those ones who doesn't have their own rides or the person who doesn't have to take their rides at the location. It has also reduced the amount of emission that happened due to many cars leaving, due to sharing facility options in this application which result in less amount of car leaving [8].

One such application of AI in road transportation is autonomous motor vehicles or in short known as AMVs. AMVs takes the help of AI to for providing self-driving cars which doesn't need any assistance from human but rather than they work on their own and carry passengers to their locations. AMVs not only can be applied in personal transport vehicles but also it could help public transport where it can run without any driver and carry numerous amounts of persons at the same time. Which will lead to less amount of human effort and can prevent unusual accident happened due to the fault of human driver and can reduce amount of emission and help in preventing Global Warming. Testing are being done of AMVs on public road and they are going with any stages of cases that they need to pass before they could travel on road by their own. One being AMVs in metropolitan cities where there is a large amount of traffic and roads are congested as compared to free-flowing highways. There is still testing that need to be done for monitoring AI behavior on these streets and they work on these congested streets.

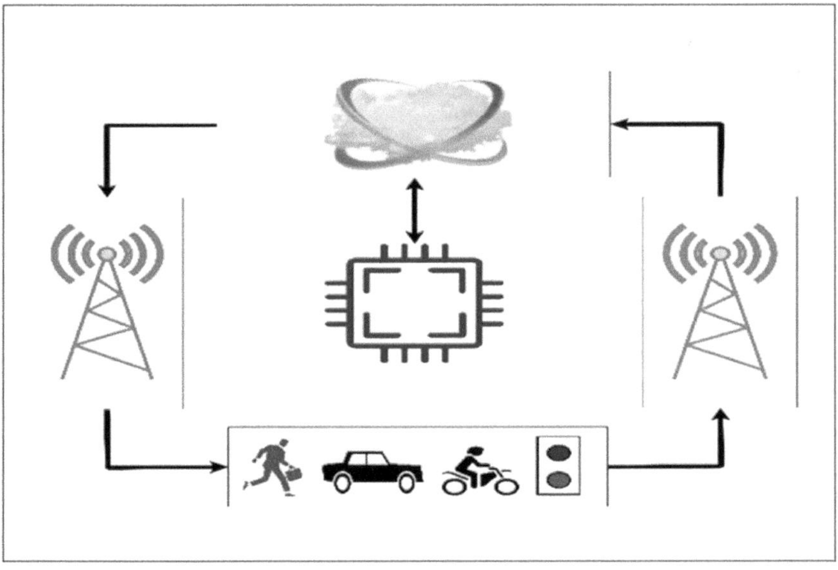

FIGURE 1.2 Artificial intelligence in road control.

Providing routs and ways to vehicles to reach their destination is one of the most used and underrated applications in Road Transport. They work on the principle of AI where they provide their assistance to human driver to reach their location at least amount of time by preventing traffic congestion and providing best route to travel on. Their aim is simple provide the best path by preventing traffic congestions which can be collected by them through internet where other AI devices sent information to. They compile data together and reroute the driver to ease them to reach their locations as quickly as possible.

AI can also help in Traffic Monitoring, by the help of machine learning (ML) internet of things (IoT) and many more. There are cameras set up on roads and highways at constant distance to monitor the flowing of traffic. In this data is collected by camera and send to clouds where AI works on Filtering things and track every movement of the vehicles. This could help in tracking vehicles, person through image recognition feature and help authority to reach to them without wasting any amount of time. They could even speed and movement of vehicles and even charge complaint directly and notify drivers about their faults so that there don't repeat the same mistake again and again. It can also help in monitoring traffic through these and dispatch authority to clear the congestion at the right time. It could also monitor the people movements, and if any suspicious thing is happening, it can alert the authority which can take action accordingly. It could also alert medical authorities if any accident happens, AI is very useful in terms of providing security at any time.

AI can help in pre booking of Parking spots and even help in guiding vehicles driver to the parking without wasting time and even help foot passengers and bike driver to cross roads by stopping the traffic at any time.

There are many usefulness of AI in road transportation but there are many drawbacks and preventions that need to be take care of. One being violation of privacy, is your applications on car is hacked using AI het hacked it could lead to violation of privacy and even led to monitoring of your whole activity without even your consent and even without you knowing.

Security is one of the most drawback with AI in road transpiration if any public monitoring devices or authority database get hacked, they could manipulate anyone and even track anyone accordingly. This could lead to mass accidents and even violation of everyone's rights. It could lead to manipulation of records and even lead to tampering with data in some cases and for removing some important data from the databases.

Loss of Jobs is one of the most concerned topics which has affected the human and which is going to do so in the process. If AI takes on this world there will be a time where there will be very less amount of human help needed. AI is taking over the work of human and at the process making them to go to other work thus creating as world where people are not sure about if they could hold their job anymore. This is creating disturbance in this world.

Lastly the topic which is in demand is it safe to give every right to AI for driving? Should we switch to Driverless Vehicles or Not? These are very concerning topic as if we give whole rights to AI and remove from any form of support who is going to take credit if any accidents happens and there is loss of life and goods. As AMV's main aim is to let the driver relax and let AI handle everything and only taking help of passenger in case of Emergency [15]. What if the emergency arrives and there is no time to waste? This is a major concern as, if the driver is relaxing how could he take control of anything this fast and what if there is no time left to take any action. Scientists are debating on this, and they still don't have reason to blame someone specific, who they should blame car owner or company owner who made that vehicles or AI who took that action. A few days ago one of the AMVs company owner stated that their AMVs are very safe and they are testing on streets with actual humans. Few days afterward their car couldn't make right decisions and resulted in two deaths. These concerns can't be left alone, and there needs to be some law regarding this before AMVs take on road.

Once there was a survey conducted to see who should be safe in case of a car accident, where you can either save car driver or 10-foot travelers. In that survey, 9 out of 10 told to save 10-foot passengers as there are 10 lives that can be prevented at cost of one. But what will AI do in this situation or any other situation like this? We don't have answers to these as we don't know what will AI do in this situation and who would it safe at this moment, and if something like this arrives who should take blame for this.

Though AI is very useful for road transportation we should not neglect that it could not make any unnecessary mistakes and that could be harmful to the society. There is concern that AI algorithms automate ethical decision-making autonomously or there should be any human intervention in this. AI is very useful in transportation till we have it under control and monitor it continually till it could make right decisions every time.

1.4 ARTIFICIAL INTELLIGENCE (AI) IN SMART METRO STATION

UAE's authorities which are responsible for transport have encouraged a significant amount of usage of ML and AI with respect to the metro services in the country of Dubai. These new and advanced AI and ML will support crowd administration during summit hours at the majority of the stations, especially after a huge public event such as concerts and so on [19].

The intention of this project is to produce an interactive metro system with AI which satisfies the daily needs of the passengers with ease and efficiency. Furthermore, this project intents to transform the locomotive travel time for metro passengers by introducing certain patterns for the trains which will control the congestion in the schedule and other important factors. This sub-project needs an extend metro replica/model to control tours of metro trains throughout the entire days. It takes into account the tickets data, Algorithms of metro access and request, and a display screen and in-depth information about user's journeys.

Outcomes of this scheme imply that metros cater to about 40% to 80% of passengers at particular or fixed locations during peak working hours and big public events/occasions, which is exhibited in the masses and more occurring runs [21].

The demo replica of this model revealed that user's movement in the heavy busy hours will lead to intervals of congested traffic and more amount of waiting time for the users/passengers prior to boarding the metro. Fortunately, the extensive and smart use of AI and ML trimmed this blockage to 40–60%, and the standing duration to half an hour (Figure 1.3).

The AI technology added to offer particular time periods and reshaping the desire to enhance the convenience of the consumers by allowing the user's specific boarding times, which enables them to reach their respective destinations on the required time. Although, these user benefits are under investigation at the moment alongside the confirmation of their usefulness and feasibility of undertaking the project shortly [21].

Big credit to the unified technological foundation, as now the RTA is more than capable of finding the intelligent explications to the objections with respect to its services, which include things like the enhanced exclusive demand. This body-always-prioritizes consumers and tirelessly works to satisfy their requirements for the sake of their happiness.

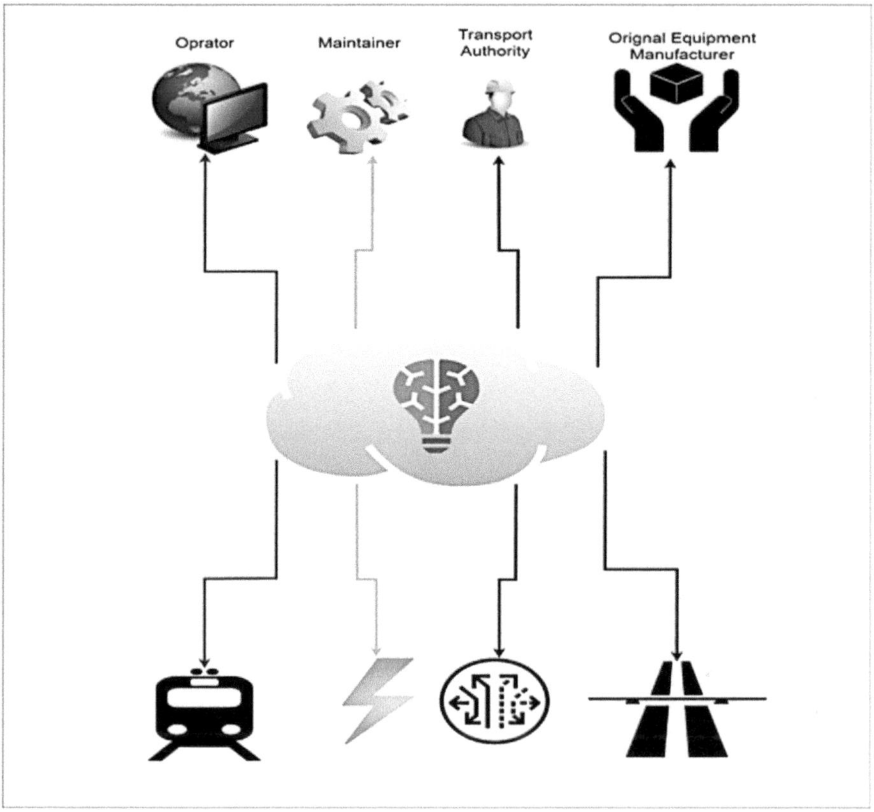

FIGURE 1.3 Artificial intelligence in smart metro system.

1.5 ARTIFICIAL INTELLIGENCE (AI) IN SMART RAIL TRANSPORT

AI and ML are intensely affecting both our secluded and professional lives, which is including our transport sector as well. Especially, the advanced and enhanced AI-driven projects and systems are hiking and strengthening the worldwide. Railway sector has been in need of huge improvements and reformations from the past 20 years or so.

Railway transportation is viewed easily after reformations with the AI because of its amiability, efficiency, and sum of money for compatibility in comparison to different ways of transportation. The responsible organization or group expects the advancement of the railways using AI and ML implemented at a high pace and time. There are some big challenges in the way of the rail industry to experience the positive effects of advancement using AI in full force, in order to overcome them, they need to extend their use of ML and AI in both the departments, i.e., railway developments and infrastructure [21]. While AI has been of concern over the past 60 or more years, only now it is starting to reach its full potential. As now we have access to advance level computing technologies which have the capability to gift us with high end ML/AI and superior UIs (user interfaces) which will ensure that these technologies are being used by as many people as possible and for their own benefits (Figure 1.4).

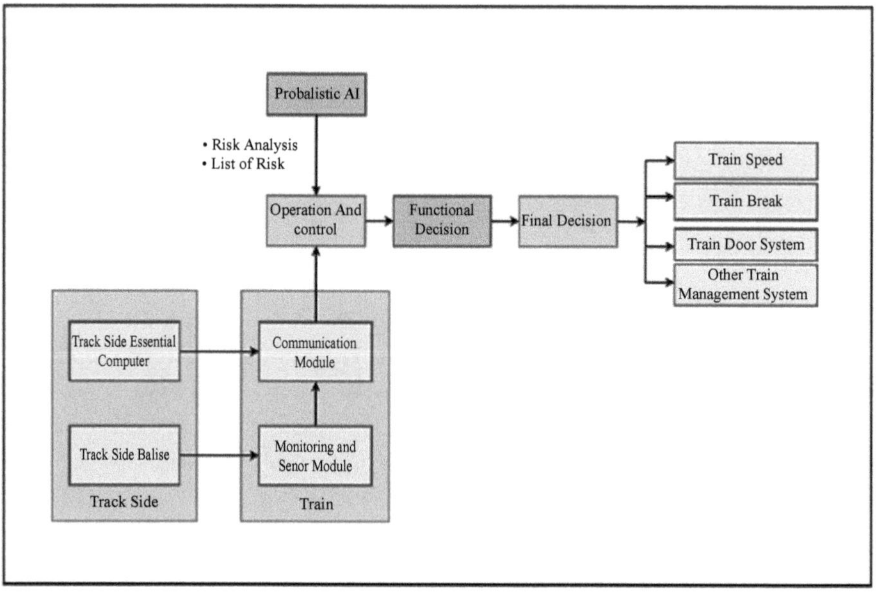

FIGURE 1.4 Artificial intelligence in train control system.

AI is enabling us to upgrade all transportation methods and make them more competent, more reliable, more intelligent, and safer. AI can help diminish traffic gridlock, recognize dangers, maintain transportation, investigate driving requirements, and even reduce arboretum gasoline effusions. Now, we are consulting about AI complimenting the railway industry to make improvements in things such as train time management and scheduling, managing the steam of a train, predicting collisions and preventing them, anticipating delays, enhancing management, and much more with an ever-increasing list of agendas. The AI advancements will improve the security of the public massively, passenger issues, and compress the all over the railway administration without compromising any of its primary functions. By implementing this method and protocol, the current system and model will work towards upgrading the trend for railway train industry to give up the share in stock markets, etc., in comparison to every other mode of transportation.

Even though there are possibly many advantages, AI also offers some complications and difficulties. AI utilization raises legal, budgetary, societal, and humane issues, for instance, what organization or person is accountable for any tragedy or loss of anything because of a cybercrime or attack or the case of integrity of information security or/and clarity. Moreover, this professes a possible menace to the masses and users, since the information can be used for monitoring which can lead to crimes indeed [12].

1.6 AUTOMATED TRAIN OPERATIONS SYSTEMS

Although AI can be implemented for the upgradation of effectiveness and diminish expenses throughout the variation of rail(ways) administration, security, surveillance, and sub-division administration department, major illustrations of the implementation of AI in rail department and area is its enrichment of train functions/operations becoming automated [12].

Automation of train operation conveys accountability for being in charge of controls with respect to the operator of the rail command system, with non-constant and inconsistent levels/degrees of independence. The IEC (International Electrotechnical Commission) has found a four conventional levels of rail mechanization: the 3rd level correlates to fully automated functions/operations which does not require a driver and the 4th level to discreet (automated) train operations. The mentioned levels are already in practice in composed train systems and metropolitan haulage systems. The Al Maktoum International airport in Dubai functions as a completely exclusive rail for the transportation of their consumers from one terminal to other using a distinguished rail track model. This utilizes

SelTrac, an automated train signaling technology that is programmed to command the rails automatically [12].

The SMRTL (Singapore Mass Rapid Transit Lines), the world's biggest automated metro system, is a very good illustration of the railway transportation system which is at 4th-level mechanization.

In Europe, the prime significant idea about the initiation of Automation of train operation and AI assistance in railways is the placement of the ERTMS (European Rail Traffic Management System), which accommodates trains with an automated driver compensation technique.

Nowadays, superior AI technologies produce a strong unit to improve and help unbar the prospective of advancement of AI-driven railways. As intermodal transportation of cars, rails, etc., is anticipated to increase, models/projects are initiated to better program vessel movements on the rail network and improve real-time data-info interchange. AI might perform a crucial part in this.

1.7 AI IN SMART TRAFFIC CONTROL

The economy of a country affects directly by arrangements done to travel road in a better way. Network management and traffic monitoring are also the issues on which the government has to focus on. Signaling system is made improperly and traffic rules are avoided. It is about traffic intrusion leading to a blockage and malfunction in traffic management resources, due to this wrong person is accused for breaking the law. The gridlock of traffic at various intersections leading to global warming and pollution causes air pollution in cities, and due to high burning of fuel, it gets very challenging for the passengers who have to wait for long hours in cars. Today's traffic management systems are unplanned and inclined to human error. The population is increasing day by day, daily passengers are increasing, there are condensed or narrow roads and global trade, due to all this, monitoring of traffic flow has become paramount importance in many countries.

Software with smart computers takes over the usual operating schemes after a technological explosion. Around the world, smart traffic management system provides improvements by providing ticket automation, strict traffic violations, safe public space, advanced solutions for traffic gridlock solutions in many major cities [13]. AI, controlled ML, computer vision, advanced IoT technology, and big data uses real-time solutions to traffic management problems. An intelligent traffic management system includes streetlights, active highways, high-resolution 3D viewing cameras, active roads and computer-assisted transportation display systems. By 2030, $25 billion is the expected reach of smart traffic management program with an 8.6% increase, according to Statistical reports. To provide better products, a lot of research is going on for smart traffic management.

The other objectives of the system are:

- Creating a standard ticket with an electronic E-bill payment system;
- Velocity sensors are used to warn passengers and drivers if speed limit increases;
- By checking real-time traffic conditions, prioritize limited traffic conditions;
- Accommodating the movement without congestion;
- Eliminate pollution;
- Big data and AI alter the performance of traffic monitoring systems and traffic control over existing synopsis;
- Advanced monitoring systems at intersections and congested roads where appropriate GPS and GIS traffic direction is required;
- Improving road communication by building an IoT, which has accelerated and improved intelligence systems;
- Use a good lighting system that stores renewable energy sources;
- Provide reliable public transportation timely;
- The big data segment and AI strengthens intelligent traffic management.

1.7.1 SMART TRAFFIC CONTROL SYSTEM

There are predefined time-based signaling method in many cities, but the shortcomings of such a method is that the traffic monitoring system works regardless of traffic flow. Therefore, the need arises for smart traffic control systems which would work according to real-time traffic conditions.

Components of smart traffic control system include smart signal lights, central control hub and AI-powered cameras and line systems. These components are defined below:

1. **Smart Signal Lights:** Smart signals and traffic lights reduce traffic gridlock inefficiency and idle time on cloverleaves. It

doesn't matter what time schedule is defined earlier, smart lights can handle the chain and free up the traffic.

2. **Central Control Hub:** Integration of central control system is done with signals, traffic radiation, chain detectors, and cameras. An AI-based system collects data from 3D AI-enabled mainframe cameras and queue viewing locations and thus, tests real-time data. For the free movement of traffic, this AI system helps to transmit an erudition clip, which is designed to control the operation of signals and traffic lights. The basis for traffic control system is the central control system (Figure 1.5).
3. **AI-Powered Cameras and Line Systems:** Line system and cameras update information about real-time traffic status on the command system. The command system then helps overcome pollution and clear congestion by sending this real-time information [3].

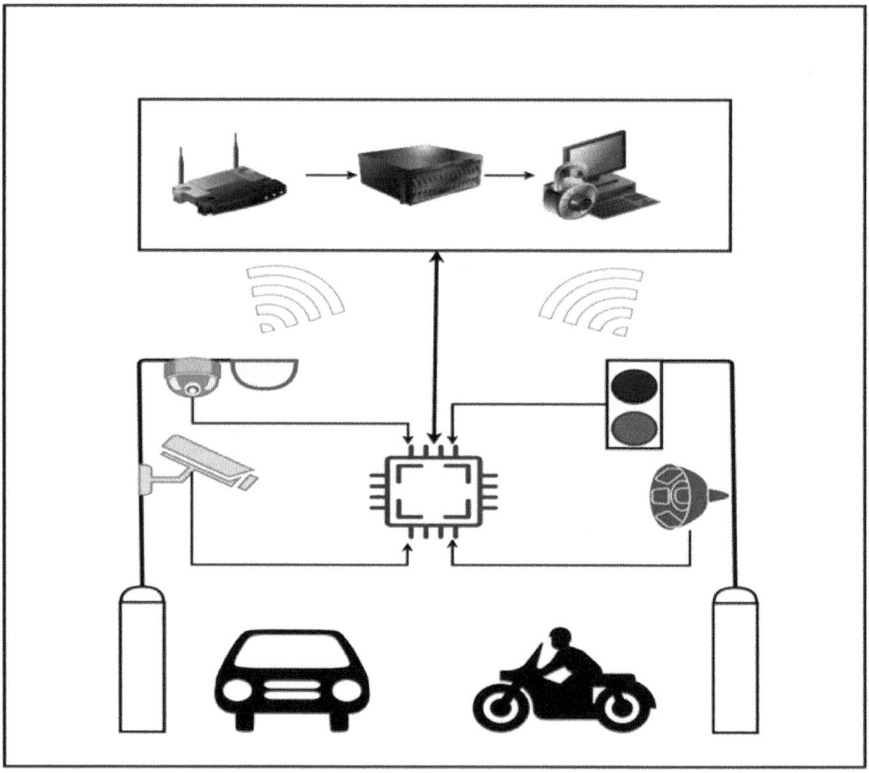

FIGURE 1.5 Smart traffic control system.

1.7.2 SMART TRAFFIC MONITORING SYSTEM

Important components of intelligent traffic monitoring systems are IoT-powered intelligent communication systems and big data tools. The technical components that define this system are smart roads, intelligent parking, traffic lamps, geospatial traffic management system and public transport. These systems change the normal functions of tickets through automation by providing superior assistance through the IoT-enabled information. Integrated sensor services, automotive information systems, and private vehicle tracking devices for tracking location. Data is accessed through an app related to parking spaces at a specific location and this data is enabled by Smart parking for better assistance. Smart highways and roads use AI technology in an increased manner [3]. To monitor the vehicle's speed and inform car owners of the proposed information systems for vehicles on the speed limit, information regarding long distances and other categories of information, roads, and highways are equipped with sensors. Geospatial traffic control system is a combination of radiofrequency devices, GPS, and GIS. Based on physical object's location, it translates 3D rendering of real-time geographical data. GPS is currently used to find the best routes to reach specific goals in a short time based on traffic in all possible directions. By providing anticipated potential obstacles, surveillance, and traffic congestion from a variety of routes and their specific queue lengths, geospatial technology can enhance the system.

1.7.3 TRAFFIC MANAGEMENT SOLUTIONS

Traffic control lights are the only thing which is dealt with by historical traffic management systems. Algorithmic solutions are very limited in solving this problem and expert system (ES) solutions are also not the best, as they have shown valid results only for signal networks which are small. As a result, the need for management of interstate highways, emergency response, overpass system and public transport is left ignored. For the advance management of traffic, a distributed blackboard system has been developed by Georgia Tech to put to use in large urban areas. To solve vehicle control issues of individuals that appear on the data structure of blackboard various things are used by sources like frame, neural network presentations, law, and script [4]. There are problems in control, monitoring, and congestion of traffic, event management and dynamic communication which are also dealt with by the sources in the system. Then the formation of traffic management decisions is done which are evaluated by simulation later on.

Certain ATMS activities includes the following:

- Display of the above streetlights is linked;
- Metering of ramp on congested highways;
- Traffic flow modeling is predicted;
- Monitoring of traffic is done;
- The enforcement of electric regulations of traffic;
- Algorithms to detect incidents;
- Opening of reversible and reversed routes;
- Control and monitoring of road;
- Needs management;
- A collection of automatic toll payments;
- Application for emergency/police vehicles.

Advanced traffic management system (ATMS) responds to dynamic traffic conditions by effectively managing available traffic resources. All modes of transport must be included in the advanced vehicle control system (VCS) in order to provide management solution like taxis, trucks, and various other vehicles. Advanced traveler information system (ATIS) provides the traveler with access to options, information special events for any given area. The accommodation details and services will also be provided to travelers. Therefore, ATIS is considered as the support system for ATMS's construction. Traffic congestion reports, police or community safety reports, vehicle management decisions, model predictions, ship line schedules and status, truck trading, event schedules, attendees, travel, and status schedules, airport schedules and status, emergency calls, incident reports, road spots, trade traffic reports, weather reports, street sensors, video cameras, cell phones, places of interest, accommodation guide, service directory are the inputs which are included in ATIS. ATIS also provides information to travelers on the existence and availability of services, car accident warning and road sign, the best routes you can take, the best travel available, availability of parking and predicted dense areas during the day [5].

Control decisions are applied on all traffic aspects, using the database of ATIS, to manage traffic by ATMs. Environmental aspects are addressed, if any, like traffic pollution, and traffic is made easier by the decisions made. Traffic control, incident management, and driver information systems are the basic elements which are used by ATM to maintain smooth traffic. There are many methods used by ATMS to manage traffic. They include:

- Reversible routes;
- Signal lights;
- Hazard warning to vehicles;
- Flexible road signs;
- Traffic signings;
- Measuring meters; and
- Emergency vehicle signals.

To maintain the traffic flow, these subsystems must work in concert. The traffic control system is useless if it cannot alert drivers to avoid a problem area and respond to accidents [23]. Many countries are still working on to implement the basics of management systems, while many countries have already done so. This shows the deployment phase of ATMS. Many modern systems have flexible control capabilities, traffic control algorithms are not just a simple timeline for the day-to-day signal systems, they are more than that. To assess the stop times, to reduce the average queues, and to determine the negative effects on traffic, control algorithms are used, namely, SCOOT and TRANSYT. United Kingdom and China uses SCOOT system currently. However, the initial system parameter is required to be adjusted for SCOOT system to ensure that its models conform to traffic visibility. To ensure system performance, periodic updates are also required. National Cheng-Kung University in Taiwan has developed computerized systems to control dynamic traffic. There are some existing street patterns and preset tables. Based on that COMDYCS-I sets the signal time schedule in its first program. While stepwise adjustment of signal timing is done as well as pre-defined traffic models are used, to determine the signal time programs logically, by COMDYCS-III. Los Angeles Automated Traffic Surveillance and Control system is a control system which is much responsive and it is used to control signal light. Signal time system is selected by comparing actual data to available data. The future plans of ATSAC's are the development of a program that is specialized to assist in identifying non-recurring congestion and also provides assistance to reduce the traffic for various categories of non-recurring events [23]. The Sydney Coordinated Adaptive Traffic

System which is developed in Australia but distributed in Hong Kong and Singapore is one of the most developed systems in the world. Real-time traffic data is collected from local computers and then transmitted to a central regional computer. This is done by SCATS system. Then the central computer selects the signal timing plan based on the information which was collected earlier.

The parameters within these systems are optimized, however, by the central computer depending on the real-time response at the local level to optimize traffic flow. TERMINUS uses the neural model to expand the philosophy of systems to create real-time traffic flow data and to determine signal light setting that will make the flow highly efficient. It uses real-time data which is collected at the local level to determine the signal light configuration, just like SCATS do. However, in TERMINUS, different configurations are determined for the current state in real time. Thus, greater flexibility is achieved in the system because different situations that occur in everyday traffic cannot be tracked by very complex systems. Detection and notification are the main functions of an incident management system. Motorists or law enforcement agencies often report traffic incidents. Response steps and ATM detection are automatic. More systems are being tested and upgraded, however. On-board automotive information systems include Japan's Road Automobile Communication Systems and Advanced Mobile Traffic Information and Communication Systems. To show other routes and problems to drivers on in-car monitors, RACS and AMTICS use microwaves and cellular connectors to local traffic stations and also digital CD-ROM maps. Similarly, to assist motorists, network connectivity is used by European DRIVE.

1.8 ARTIFICIAL INTELLIGENCE (AI) IN FESTIVE TRAFFIC CONTROL

In urban cities, traffic is congested and overly busy; for instance, take Boston, where drivers witness an interstate gridlock in the country. The population of the England city allocated an aggregate of 174 hours of hunkering with respect to the transportation being static on the road leisurely, which led to a loss of at least 2,291 dollars. Therefore congested traffic management may cause loss of money and time both. Moscow City's straphangers lose around 210 hours per newest year because of transportation/traffic mismanagement. In retaliation to this, the responsible bodies and organizations are preparing to implement AI and ML to solve/decrease the traffic congestion and gridlocks. Similar situations of congested traffic arise when there are big occasion specifically festivals. Every city as an authentic set of challenges, but the solutions to these set of challenges can be identical. Fixing these challenges using AI and ML can be highly productive as we can predict the traffic during the festive days and actually manage it according to the traffic state which is predicted, similar to disaster management [18].

Countries are looking into AI and ML in order to solve these traffic-based problems, AI and ML have policies which are related to cloud-based technologies that can efficiently join the government, cars or any transport asset, the population and then optimize or decrease the possibility of traffic congestion by mutual co-ordination [11]. For better understanding, the cloud platforms can be compared with command towers for an army, which gives instructions for better co-ordination and avoids confusion. These levels of cloud-based AI ensure moving providers to assure transit functions/objective by presenting a huge amount of preparation, administration, protocols, foreknowledge, and market science understanding for heterogeneous transportation assets.

The entire process demands the omission of a single responsible body in the city, and this would result in third-party intervention for the good as it can mold the full system into a more responsible and sustainable plan, which is better than having the individual providers such as today's assisting brands of transports.

This festive control system needs to take real time situations that occur on non-festive days into consideration and work on it. For example, on Miami's highway bridge, car drivers and passengers have a waiting duration of 10 to 20 minutes if they get stuck in an above average traffic situation, now this can be used as a modeling question to solve festive related problems as we can give the algorithm the learning data then predict future traffic states and work accordingly.

All around the globe, practical of advanced traffic solutions are being bought into use. And while some of it has been highly focused on things such as a sustainable self-driven transport asset technology which are AI- and ML-powered platform that is leading to an unreal advancement in transport transit sector.

These assets are being used for self-driven, human-driven, in relation to municipal corporation and individual ones in a connected method of main tower which was mentioned earlier. With enhancements to this, we can actually use these AI-based assets for festive traffic control, for instance, we can use these self-driven assets for some extra work during festival days which will actually save our personal time and space during those particular days and there are other practical usages of these assets as well.

In this method of implementation, the AI and ML-inspired platform allows vehicular assets to plan, establish, and control the management of services using any class of vehicular asset. This allows fluidity among the providers to provide the deliverables that transporters require, such as hailing rides, robo-taxi, micro-transit and shuttles which are independent – every one of these helps the masses to travel from position D to E, in possibly in the utmost efficient method or way [18]. This method can also be implemented to deal with festival traffic management but it might require patience and research budget with a little cooperation from the governing bodies.

Alongside the implementation of AI and ML to enhance the program of festive control in order for it continually grow/improve its ability/ capacity and efficiency, to shape this program into something which is revolutionary in the transport industry. Therefore encouraging manual and automated vehicles will lead to their usage in day-to-day life in the coming years and not only in festival traffic control.

After all these powerful high-end AI and cloud-based ideas finally come together. Then, they will lead to exceptional traffic management with less labor and more efficiency. As it will provide assistance easily based on user need and demands, that too not only on festive days but in our daily lives as well. Moreover, self-driven cars might actually reduce congestion and accidents which happen more often than not on festivals. All over the potential is too high but we just need to be patient with the ideas [18].

1.9 USES OF ARTIFICIAL INTELLIGENCE (AI) IN NATURAL DISASTER TRAFFIC CONTROL

All over the planet, natural calamities have been a regular and normal occurrence. These disasters have been increasing over the years for a lot of reasons. In the year 2017, typhoons, quakes, and forest fires caused a loss of $306 billion globally, which is around 2-fold than 2016's loss which was $188 billion.

These calamities are a consequence of weather variation, violent weather, aging inadequately shaped foundation, amongst various other problems, there is a problem which relates to the well-being of us Humans and our lives, due to natural disasters. Universally, 94 trillion dollars investment is required to maintain the speed which corresponds community progress as well, with a huge part of it invested in the restoration of our natural habitat. That being said, a large amount must be invested in the well-being and traffic organization during disasters so loss of human life and vehicles can be reduced as every year in traffic accidents caused by natural disasters is a lot and we can use AI for this purpose.

One of the simple explanations can be understood with the help of this given diagram, here it implies that we can predict the state of the traffic and manage the traffic accordingly; similar to festive traffic management, using deep learning and neural networks. The basic idea is there but we need significant amount of data for a good sustainable algorithm. Moreover, the IOT, AI, and ML are some of the most powerful helping tools for this scenario. These powerful assets can develop intelligent predictions and decrease the individual cost along with the foundation cost of natural calamities when they actually occur. Catastrophe modeling is a great initiative which can lead to significant advanced inclusive programs in order to tackle calamities and solve the issues related to them very efficiently [17].

1.9.1 ANTICIPATING DISASTERS WITH BETTER DATA

Fortuitously, the technologies and the process of predicting natural calamities keep on evolving into advanced and more accurate systems, allowing national tech bodies like FEMA, NASA, and NOAA to design protocols and solutions. These bodies utilize data from various sources such as satellite pictures, sensors, and so on for predicting natural calamities and find solutions to them in prior. An instance, where natural emergency department can easily look into the situation using AI and cloud-based software which can alarm the towers and the responsible authorities for calamities such as typhoons and tropical disasters. Moreover, data can be collected and used using sensors on trees in case of a draft which is highly common in some parts of the world. One more example is SkyAlert, a high-end functional application which is used in Mexico for the same purpose. It takes advantage of mobiles, computers, and other such devices, and there is one more great example which uses IoT technology on

Microsoft's cloud platform Azure for alarming millions of people before an earthquake or any such thing is about to occur, which ultimately allows the users to escape and move away quickly before the disaster hits [17].

1.9.2 PREPARATION AT A GRANULAR LEVEL

Advanced technologies such as AI and IoT are installed to capture data and real time examples from highways, roads, buildings, and then this data is sent to an organization that takes out meaningful information from the captured data and then uses it to improve the general experience [18]. These operational technologies can make be a great defensive barrier and provide fast instantaneous services to the people during a natural disaster without much panicking and loss of manual labor. When compared to some of the current working models which are based on analog signaling, these AI-based models are an amazing upgrade on the current system and will only continue to grow.

Councils on smaller level which represent a town or a small city can also access AI and IoT to predict natural calamities in their area, and warn the population about a tackle plan which will keep them safe by warning and evacuating the area which is to be affected. During a real life disaster scenario AI and ML could easily come in handy by constantly notifying the masses about safe routes by which they can escape or avoid the damage using relative position of each individual or group. AI can tell which route to take without causing any congestion or confusion if the population is willing to understand its working like WhatsApp. Here is one good example of evacuation using relative positioning. If there is a blast or a natural calamity in stadium or a shopping mall then AI-based models can actually help people find the best exits along with which ones to avoid which will be a huge step forward in this area.

1.9.3 RESPONDING MORE EFFICIENTLY

The three days right after any natural calamity are very crucial. Different set of organizations or concerned bodies should come up with protocols, establish functions to aid, look for any remaining human which can be saved or aided, and initiate programs to make sure that there are no or least possible environmental variations due to the calamity. AI and internet of things (IoT) can massively help in providing in depth insights which can be used for improving disaster traffic, such as [19]:

- Images from aerial visions and space devices under government;
- Internet of things and AI base data;
- Artificial intelligence-powered chatbots;
- Social media in relation to cloud.

The following ways will assist the responsible authorities to anticipate required aiding materials, plan evacuation protocols, and avoid unimportant steps which will consume time and sources, but only if it's decipherable. AI is capable of detecting calamities prior to their actual arrival if they are fed with the data of past disaster and with time new data can be generated too so there is only improvements which are going to take place with respect to AI predictive models for natural disasters. Same can be said for ML-based models which can predict any type of disaster with great accuracy [20].

For instance, a big public building is affected by a natural calamity, and they have AI-based help available to them. Using the AI-based the safety team can come with an aid plan for the people stuck or affected which will take the design and the whole body of the affected building in consideration and act accordingly, scan the nearby areas and see how many people require aid which will give the helping team a better idea. It can also figure out the affected transportation assets and help them to show a better way to get out of the red zone area by using google maps or something similar, last but not the least it can help transportation during or after the disaster which will lead to less congestion and a smoother way to keep everyone secure. Using AI and IoT with cloud-based models which uses devices such as mobile can lead to faster traffic management during disasters. In future, when we have enough knowledge and resources, a humanoid based on AI is possible too which will provide fast and save care in congested traffic situations due to calamities [18].

1.9.4 PROGRESSED PLANNING AND ASSISTANCE EFFORTS THROUGH ANALYTICS

AI and ML will be very important in providing assistance to traffic officials as well, it will allow them to help the affected and needy in case of disaster on priority bases and will decrease the threshold time for their assistance, proving to be essential for both of the parties. AI can be implemented to search or predict the most affected (to be) areas by a natural disaster by taking several parameters into account such as season history, disaster history, usual traffic trends and many others by doing this the departments can prepare a plan to help the vehicles in that area and an emergency blueprint as well. These findings and data can also be used to make less prone disaster buildings or highways in areas which are predicted to be affected more, this way natural property damage can be easily avoided [19].

Some disasters are inevitable but most of them can be avoided on road or at least the impact can be minimized, as we have the power of AI to prevent and prejudice the disasters before they becoming a bigger problem, as now the historical data will prove to be essential when dealing with transport problems and we can prepare and manage an evacuation plan for the road traffic, which will lower the number of victims and lower the loss of transport assets [19].

In the times of natural disasters, AI will act as a supervisor and a very responsible one because the potential of AI depends upon the data to some extent and we have enough sources of data to create an ultimate disaster traffic manger based on AI, in near future.

1.10 TRAFFIC CONTROL IN RELATION TO SOCIAL MEDIA

At this time, various cities of all over the world are facing lot of problems related to the traffic congestion. The increasing of vehicles, either personal or public transport are the big issue for controlling during the jam. Because of this, the earnings of the countries are reducing. This reason is creating the huge time spent during the traffic jam. However, the huge risk is appeared with traffic jams include vehicle break-ins during the traffic jams.

According to the aforesaid reason, roughly 5 billion unit of electric current and 2 billion gallons of burning products are wasted annually during the traffic jams.

The emergency services like police and medical-related services in the cities, also increasing the vehicles conflicting together by the drivers rashly and avoiding the traffic rules. The congestion of the road is another reason due to increasing the quantity of vehicles. According to this, road is not maintained, transport is not properly planned, and public transport is also not managed, all this creates a huge problem for traffic jams and because of improper traffic management, people are facing big problems for this situation in the cities. In order to manage the traffic jams, the actual time of events occurring on roads would be known to people. There are two types of traffic congestion such as recurrent and non-recurrent.

1.10.1 RECURRENT TYPE OF TRAFFIC

This refers to the use of limited space of road every time by a wide range of vehicles.

1.10.2 NON-RECURRENT TYPE OF TRAFFIC

This refers to events like weather situations, accidents, vehicles break down all these are random and unplanned and day off celebrations, etc.

According to this, non-recurrent traffic is compounded by the recurrent type of event. Therefore, traffic lights, which is a conventional methods of traffic control are not working on the various roads where applicable. However, the unfavorable effects appeared on the smooth functioning of traffic. As per the situation, the traffic officer should control and manage the traffic so that we can solve the problem during the traffic jam [7]. Now, the Social Network is important in our daily lives. Data is combined along with the picture or video for good explanation in any situation by social media. It provides real time information which is backed by research.

Traffic congestion's impact should be lessened and policy choices should be informed. If any similar crises occur, it's copy should be done to fight against that crises. Road traffic administration is influenced both extensively and intensively by social media. Roads that are closed, accidents, and other similar situations on the roads are also accomplished by social media [7].

To manage the huge blockages on road traffic, social media's opinions and suggestions are utilized. Road jam in cities has triggered the social media criticism, it is being said by literary suggestions. The activation of this discussion is clear that this information is developed and achieved by the network. Individual are also active on multiple groups on various platforms. Hence, Facebook is the most popular platform all over the world for guiding and controlling the road traffic administration. The other social media platform is Twitter, which has played a very important part for controlling and guiding the road traffic management in the cities. According to this, different policies are adopted by different countries which is related to control the road traffic problems. Therefore, small groups are the basic components of social media in which members experience various choices. These non-recurrent traffic leads to not only in one-to-one conversation criticism of authority but also on various networking sites and social media handles [9].

1.11 AUTONOMOUS TRANSPORTATION VEHICLE

Automated vehicles is next big thing in transportation that is happening and are in ongoing phase. Automated vehicles testing is happening around the world, even some of them operating on freeways on their own. Arrival of ATVs on roadways is going to impact the way we look at transport and it is going to create a huge turnover within whole transportation [21]. ATVs will create a new era where command of the vehicles will be shifted from user to the AI, which will be the first time where AI takes control of full vehicle and human are just there for the show. Working of ATVs is very simple with help of AI sensors, applications, trackers they will monitor the surrounding and move accordingly. They could even react to sudden movement change by any other cars and can even change course accordingly. AI is playing a huge role in ATVs without AI ATVs could not operate on roads. Many private companies and working on making the first secure ATV which could operate on roads with ease. Many companies are testing their ATVs on freeways, and some are even ready to get to operate on metropolitan streets. If they successfully get their testing done on only, they will take the whole transport system which storm but they could generate uncountable revenue. ATVs not only affect the private sector but it will also create a huge impact on the public sector also. With its coming to the public sector, it could take over not only Road transportation but also airways and seaways also. There is no denying that AI in ATVs will react fast, will be more responsible, more effective, more convenient, and more economical. It can react fast as compared to human senses also [10].

But there are a few concerns that are taking place with ATVs and we don't have any solution for these concerns. First being what if AI does some mistake and it leads to some serious accident and even damage to property and lives, who will take the blame for this, person who owns the vehicle, company that made that vehicle, AI who was operating that vehicle. As main aim of ATVs is to operate without human interaction and only consult human when necessary, what happen if human could not take over at the given time, which is mostly possible as tested you can't handle a huge strain in one second when you are resting a second before, it takes some time for human mind to react, so we can't blame user for anything and even company whose cars were released to operate after full testing and government approval. And in case of AI, you can't blame or accuse a non-living thing. This concern is huge and as there is no law regarding AI, there is no assurance we are save if something like this happens. So, there should be some rules and regulations that cover AI involvement in ATVs [22].

1.12 RISK AND CHALLENGES

No doubt AI can significantly advance efficiency and productivity, but it can also have social and economic consequences that need to be managed. Some of the most significant consequences are mentioned below:

1. **Job Loses:** According to a report by the Center for Global Policy Solutions, it is seen that in a transition to AVs in the United States, more than 4 million jobs would likely be lost which include bus drivers, taxi drivers, etc. Ai can accelerate job losses for low-skilled workers in fields like transportation by boosting the development of economic models which are already established as it made a transition toward a service economy.
2. **Regulatory Requirements:** When AI systems act wrong just like humans do, then it is hard to foretell what should the regulatory requirements be. It is not easy to understand who would likely be responsible if AV would do any mishap or injury in spite of the fact that according to a research, AVs lessen deaths caused by traffic.
3. **Privacy Concerns:** There must be strong privacy laws for ML where users are asked to opt and provide personal data. Too much information on social network should be balanced with the privacy laws.

4. **Lack of Skills and Education:** In developed countries, investment in AI is growing and consequently over the last few years AI experts are in demand but the greatest barrier in adoption of AI is scarcity of skilled AI people. AI is a complex technology with social and economic effects which is why it is difficult for a developed country to successfully make use of it. In order to get most of the benefit from AI, it takes time for a developed country to have skilled people and to build effective business strategies.
5. **Cost:** In the transport industry, it is seen that AI is not progressing well. The main reason is that AI systems are overpriced. Applications of AI in the transport industry is not fully utilized by inexpensive and not fully developed countries since they are not ready yet to implement AI systems and also to provide services to these systems. To fully utilize the power of AI systems in transportation, countries must be investing in AI and those countries which do not invest in these systems may have difficult time to fully utilize AI power.

1.13 CONCLUSION

AI and its subset, such as ML and so on, has the potential to exponentially boost the positive output and profit in a lot of public and private sectors, which includes transportation as well, and these changes are not as far as some might anticipate, one can observe and see these changes at this very moment in their daily lives. AI is making transport safer, quicker, and more convenient for normal public right now and it will only continue to grow. For instance, AI-driven drones are delivering medicines and other essentials in places such as Afghanistan and rural Africa, and there are AI-driven traffic management systems that decrease the emission and congestion of traffic and pollution in Japan, and automated cars like tesla and the cargo shuttle in China, some of these do not even need a driver that is how advanced they are. With tremendous ability to increase performance and durability, along with more positives, there are a few negatives which relates to social, demographical, industrial, and political challenges that must be solved in order for peaceful use of the power of AI evolution and progress in every sector of society and universal benefits of the general public, which includes transportation as well [14].

KEYWORDS

- **advanced traveler information system**
- **artificial intelligence**
- **autonomous motor vehicles**
- **European rail traffic management system**
- **International Electrotechnical Commission**
- **Singapore mass rapid transit lines**

REFERENCES

1. Zhang, Y., Wang, G., Gu, X., Zhang, S., & Hu, J., (2015). Real-time pedestrian detection with the videos of car camera. *Advances in Mechanical Engineering, 7*(12), 1687814015622903.
2. Singh, A. P., & Gupta, M., (2021). Video based vehicle and pedestrian detection. *Annals of the Romanian Society for Cell Biology, 25*(6), 14653–14658.
3. Papageorgiou, C., & Poggio, T., (1999). Trainable pedestrian detection. In: *Proceedings 1999 International Conference on Image Processing (Cat. 99CH36348)* (Vol. 4, pp. 35–39). IEEE.
4. Bertozzi, M., Bombini, L., Cerri, P., Medici, P., Antonello, P. C., & Miglietta, M., (2008). Obstacle detection and classification fusing radar and vision. In: *2008 IEEE Intelligent Vehicles Symposium* (pp. 608–613). IEEE.
5. Qi, B., John, V., Liu, Z., & Mita, S. (2016). Pedestrian detection from thermal images: A sparse representation based approach. *Infrared Physics & Technology, 76*, 157–167.
6. Arras, K. O., Mozos, Ó. M., & Burgard, W., (2007). Using boosted features for detection of people in 2D range scans. In: *Proc. of the IEEE Intl. Conf. on Robotics and Automation.*
7. Zeng, K., Liu, W., Wang, X., & Chen, S., (2013). Traffic congestion and social media in China. *IEEE Intelligent Systems, 28*(1), 72–77.
8. Sullivan, J., & Carlsson, S., (2002). Recognizing and tracking human action. In: *European Conference on Computer Vision* (pp. 629–644). Springer, Berlin, Heidelberg.
9. Juhlin, O., (2011). Social media on the road: Mobile technologies and future traffic research. *IEEE Multimedia, 18*(1), 8–10.
10. Lv, Y., Duan, Y., Kang, W., Li, Z., & Wang, F. Y., (2014). Traffic flow prediction with big data: A deep learning approach. *IEEE Transactions on Intelligent Transportation Systems, 16*(2), 865–873.
11. Claussmann, L., Revilloud, M., Gruyer, D., & Glaser, S., (2019). A review of motion planning for highway autonomous driving. *IEEE Transactions on Intelligent Transportation Systems, 21*(5), 1826–1848.

12. Hodge, V. J., O'Keefe, S., Weeks, M., & Moulds, A., (2014). Wireless sensor networks for condition monitoring in the railway industry: A survey. *IEEE Transactions on Intelligent Transportation Systems, 16*(3), 1088–1106.
13. Veres, M., & Moussa, M., (2019). Deep learning for intelligent transportation systems: A survey of emerging trends. *IEEE Transactions on Intelligent Transportation Systems, 21*(8), 3152–3168.
14. Eskandarian, A., Wu, C., & Sun, C., (2019). Research advances and challenges of autonomous and connected ground vehicles. *IEEE Transactions on Intelligent Transportation Systems.*
15. Claussmann, L., Revilloud, M., Gruyer, D., & Glaser, S., (2019). A review of motion planning for highway autonomous driving. *IEEE Transactions on Intelligent Transportation Systems, 21*(5), 1826–1848.
16. Cui, Z., Henrickson, K., Ke, R., & Wang, Y., (2019). Traffic graph convolutional recurrent neural network: A deep learning framework for network-scale traffic learning and forecasting. *IEEE Transactions on Intelligent Transportation Systems, 21*(11), 4883–4894.
17. Jiang, G., Meng, Z., Wu, Z., Li, Q., & Zhang, C., (2010). Traffic adaptive control framework for real time large-scale emergency evacuation. In: *2010 International Conference on Intelligent Computation Technology and Automation* (Vol. 3, pp. 1065–1068). IEEE.
18. He, M., (2010). Study on emergency management of traffic jam and improvement of emergency mechanism during ice disaster in human. In: *2010 IEEE International Conference on Emergency Management and Management Sciences* (pp. 318–321). IEEE.
19. Sun, X., Yang, Z., & Wang, Z., (2010). Multi-level hierarchical layout method of large scale subgrade-type traffic detectors under disaster. In: *2010 International Conference on Intelligent Computation Technology and Automation* (Vol. 2, pp. 748–751). IEEE.
20. Treethidtaphat, W., Pattara-Atikom, W., & Khaimook, S., (2017). Bus arrival time prediction at any distance of bus route using deep neural network model. In: *2017 IEEE 20th International Conference on Intelligent Transportation Systems (ITSC)* (pp. 988–992). IEEE.
21. Kumar, D. R., Krishna, T. A., & Wahi, A., (2018). Health monitoring framework for in time recognition of pulmonary embolism using the internet of things. *Journal of Computational and Theoretical Nanoscience, 15*(5), 1598–1602.
22. He, Y., Zhao, Y., & Tsui, K. L., (2018). An analysis of factors influencing metro station ridership: Insights from Taipei metro. In: *2018 21st International Conference on Intelligent Transportation Systems (ITSC)* (pp. 1598–1603). IEEE.
23. Desai, Y., Rungta, Y., & Reshamwala, P., (2020). Automatic traffic management and surveillance system. In: *2020 International Conference on Smart Innovations in Design, Environment, Management, Planning and Computing (ICSIDEMPC)* (pp. 131–133). IEE.

CHAPTER 2

AI Approaches in Intelligent Transportation Systems

K. NITHYA,[1] S. MYTHILI,[2] M. KALAMANI,[3] and M. KRISHNAMOORTHI[4]

[1]*Department of Computer Science and Engineering, Kongu Engineering College, Tamil Nadu, India*

[2]*Department of Electronics and Communication Engineering, Bannari Amman Institute of Technology, Tamil Nadu, India*

[3]*Department of Electronics and Communication Engineering, KPR Institute of Engineering and Technology, Tamil Nadu, India*

[4]*Department of Information Technology, Dr. N.G.P. Institute of Technology, Tamil Nadu, India*

ABSTRACT

Modern society is reacting according to the decision of intelligence supports. Two such influencing technologies are IoT and AI. Most of the lives and the valuable times of humans are saved using the intelligent transportation systems (ITS). This chapter covers various functionalities of IoT-enabled transportation systems using smart devices like sensors and its usage in three verticals such as on road sensors to control the traffic and efficient parking, in vehicle sensors to prevent accidents and on body sensors to monitor the driver's condition. And then it is broadened towards the AI functionalities in ITS where the IoT network is deployed. The focus on safety, traffic management, parking management, electronic toll collection, automatic fare collection, etc., are bringing up the cities and

Artificial Intelligence for Future Intelligent Transportation: Smarter and Greener Infrastructure Design. Rajesh Kumar Dhanaraj, Nilayam Kumar Kamila, Subhendu Kumar Pani, Balamurugan Balusamy, & Vani Rajasekar (Eds.)

its transport facilities on the roadside in a better way. Speed alert system, emergency vehicle notification, driver drowsiness detection, etc., are the supportive systems to ensure public safety in all aspects. As an exemplar, a model is implemented and tested using a kind of on body sensor as eye blink sensor to detect driver's drowsiness in a smart way. Further, with the output of the sensor an AI technique as CNN algorithm is used to classify and decide the feel of the driver accurately. This helps to take the further action to alert the driver by its decision. This implementation ensures safety for the passengers and prevents accidents.

2.1 INTRODUCTION

Entire world is getting revolutionized with the help of smart technology and its various advancements. The real time scenarios are being reflected in implementations of smart city projects like transportation, healthcare systems, infrastructure, accessibility, etc., among all, the transport Management system plays an important role in developing smart cities. As per the Ministry of Land, Infrastructure and Transport, the government policy experiences more inclusive growth in the construction of bridges and roads to make the transportation in an intelligent way which directly supports the country's economic growth and globalization. In another perspective it reflects in the employment opportunities involved in the transport sector such as car, taxi, van, and bus drivers, conductors, railroad workers, road construction labors, supervisor, public transportation inspector, streetcar operator, logistician, traffic supervisor, transportation manager, transportation planner, travel coordinator, tourism, etc. In the developing nation, mobility is required in high aspects for the people to travel for daily work, shopping, and for many social reasons. In general, transport is of two categories: private and public. To travel from one location to another location to reach the destination the mode of transport will be preferred by the passengers that maybe bicycle, car, bike, bus, train, etc. Due to the economic growth dependency of transportation, the location accessible cities will show higher the cost for transportation due to the huge traffic which limits the welfare of the society thus in terms of air pollution and using large amounts of land to travel. Detained issues such as robbery, pickpocket, mass killing which is a part of terrorism effects all together threatens the people who are in travel in order to contribute for these circumstances and enhance the security level smart transportation systems (STS) will be recommended. Around 60% of the people are living in cities and it will be increased to 65% by 2040. Around 1.3 million people who are in cities are driving their vehicles and roaming on the roadside area, with the support of wireless network radio technology and sensors all together brings many smart city projects and its implementation. The overall population is getting shortened due to the traffic crash on the road and it reflects the cost of 3% in the country's GDP. It has to be taken care of to manage the natural environment and reduce accidents. There should be a realistic solution. One such supportive solution is the Intelligent Transportation System.

2.2 OVERVIEW OF ITS

Fast computing for vast operations is essential in day-to-day life. Today's communication technology has proven intelligence in transportation with respect to safety and smart moves. The roadway accidents are getting reduced and provide a smart way to make a move of transportation even in traffic areas [2]. Vehicles are moving like water flow because of smart transport systems and traffic management. Operating and managing the transportation system's safety and security is a big challenge but still achievable in the fast-growing number of vehicles world. By managing the traffic on the road and due to the toll gate, parking systems are extremely useful to avoid high-speed traveling for the case of emergency time to report. Traffic system structure and its efficiency has to be concentrated if there is an explosive number of four wheelers and two wheelers on the road. If that happens then the speed of vehicle movement is increasing from 15% to 20% and there is a possibility for fuel consumption also. It further decreases the death rate. Severe injuries due to the road crash is mainly due to the mobile phone's usage, not wearing the helmets, over speed, etc. As per the global status report the, road safety is based on road infrastructure and traffic management system.

2.2.1 RECENT STUDY ON ITS IMPLEMENTATION

1. The entire public bus information is available to the passengers who want to get into the bus the information such as bus timings, current location, seat availability, etc. It is in the city of Glasgow belongs to Scotland.

2. The recent design in which the routes of night bus services will be addressed based on the mobile phone location of the accessible users and provide the service based on the level of demand. It has been launched by the Seoul government.
3. Optimized ambulance location based on the emergency call pattern in the city of New Orleans.
4. Driverless cars arrive on streets to reach the destined location autonomously planned and in progress by Sydney metro.

2.2.2 APPLICATIONS OF ITS

The growing number of population and the use of vehicles on the road makes slower speeds, long trip times, and increased vehicular queueing. As a whole the substantial increase of vehicles causes extreme traffic congestion. In the urban area, the predominant road facilities are the case for traffic collision for any mode of transport. ITS is a platform that supports the enhancement of quality and reaches the outcome in a controllable manner. Some of the applications of intelligent transportation systems (ITS) are efficient parking barrier management, electronic toll collection, emergency vehicle notification system, dynamic traffic light sequence, collision avoidance system, speed alerts, and traffic management. The information and communication technology like the internet of things (IoT) paves a way for efficient traffic management through the sensors where the real time data is fetched, processed, and transmitted. Further, it leads to avoid accidents, autonomous parking space identification in both rural and urban areas and also optimize the use of public city transport.

The most important application of intelligent transport systems are traffic control and management. There are many reasons for the traffic congestion. The traffic incidents such as work zones, special occasions, weather conditions are the primary factor that affects people's time.

2.2.2.1 TRAFFIC CONTROL AND MANAGEMENT

The crucial aspect of traffic management enhances the quality of a particular city. So, the basic infrastructure problem is resolved by the construction planning and implementation. The success of IoT and its machine design for traffic control is purely about Sensors and cloud-based platforms. To examine, gather, and mine the collected data is possible with IoT devices. Mohammed et al. [5] proposed an IoT-based system model in which the real-time traffic scenario is observed. The system architecture consists of four layers, namely sensing layer, network layer, service layer and application layer. The sensing layer which consists of sensors and many active devices will sense the vehicular data and the wi-fi-based microcontroller installed will forward the data to the service layer. The open-source IoT platform is used to the devices which are connected for communication via the process of storing and managing. The device system is installed on road intersections which holds message display units to deliver the traffic message to the end user then and there. The unusual road incidents are also intimated with the expectations of road clearance or alternate route suggestions.

2.2.2.2 EMERGENCY VEHICLE NOTIFICATION

An approach of notifying the traffic scenario for the emergency vehicle is mounted with several alert units on a roadway [6]. The signal emitted from the transmitting unit of the emergency vehicle like ambulance is received by the detector unit on a roadway it compares the received signal with the proximity threshold of that exceeds it further notified the non-emergency vehicles through the light glowness. Thus, this makes the other vehicles know about the emergency vehicle which is on the road. The light flashes towards the emergency vehicle direction. The communication is continued by transferring the alert signal to the next alert unit on a roadway and the relay unit. This kind of prior information about the vehicles via signal propagation helps to avoid the death rate. The alertness given for the drivers through this technique helps to clear the path and save the lives of the humans who have got stuck up in the traffic even in emergency conditions. The equipped lights and sirens employed on the roadways will help for vehicular navigation on time by alerting the non-emergency vehicles. Another approach is e-call, the autonomous location and position of the vehicle is notified to the emergency help centers in European countries. It is made possible with the onboard sensors like air bag sensors which are active when accidents are met. There is a push button which helps to make a call manually by the occupant inside the car using 112. The trapped location is identified using the satellite while the call is made. With this data, the trashed location will be reached by the rescue team to save the people in the vehicle. It is proven that the death rate of 4–8% is reduced in Finland using this approach.

2.2.2.3 EFFICIENT PARKING MANAGEMENT SYSTEM

According to a survey report the 30% of traffic is due to the time taken by the drivers and motorists to find the spot for parking the vehicles. Most of the people consider this as the most frustrating activity and time consuming while dealing with the vehicle in urban cities. It can be made autonomous and easy with the help of an intelligent parking system which obviously falls under the category of intelligent transportation system [4]. The sensors and cameras are the major instruments to gather data in real time to measure parking occupancy and its status. Artificial Intelligence (AI) provision is for analyzing a wide variety of data like frequency of future vehicular traffic, Traffic in peak hours, type of vehicle, etc. The services like Advanced navigation services, knowledge about the traveling location, Traffic scenario, etc., and its data updating leads to the simplified functionalities which can be achieved with the help of AI. One of the AI-based systems is ALPR (automated license plate recognition) [9] which detects the number plate of the vehicles using 360° camera and optical character recognition patterns and then automatically frees up the space for parking of that vehicle without manual intervention. It is user friendly and in vehicle application software which consolidates the information of where the car is and how long it has been there and all other details. This detailed data of vehicular parking helps to find the vehicle if someone steals it. The major advantage through this is that the inconvenience of commuters gets eliminated, Security enforcement, Reduction of Fuel consumption through avoiding the unnecessary wastage of time while finding the free space and it supports the parking operator to provide efficient space for the incoming vehicles. Human assistance along with machine intelligence provides the efficient parking management system. The future focus towards the parking management is using AI-based robots for parking the vehicles if the user closes to the destination. It is a proven solution to limit the operating cost and the human workload of parking managers, parking patrol and the parking users even as the expanded technology is in use.

The IoT devices and several sensors help for the cloud-based location information updation. The IoT devices like Raspberry Pi, Arduino for parking occupancy detection using camera-based IoT systems. The traffic surveillance is done with the help of cloud data processing using several algorithms in an online manner for transmitting the data to the server database for parking allotment when can be in street and off street. For fast computing, multiple sensors like magnetic, ultrasonic, infrared, and loop sensors are used to detect multiple parking spaces. The parking occupancy status is efficiently determined using the IoT-based system.

2.2.2.3.1 Parking Management

Uses various sensors on the road to minimize the traffic via intimating the real time situation to the vehicles using radio communication. In this way, the parking lot situation is also known prior using sensory network communication [4].

2.2.2.4 ELECTRONIC TOLL COLLECTION SYSTEM

Electronic toll collection System is the fastest alternative to manual toll collection [3]; It collects toll fees automatically when the vehicles cross the toll road, toll bridge or toll tunnel. In traditional toll collection methods, the driver needs to stop and pay the toll fee manually by cash or card. Electronic toll collection performs the various automated tasks in a sequential way; they are vehicle detection, vehicle classification, transaction processing and violation enforcement. Automated vehicle detection systems identify the passages of vehicles in toll. Once the vehicle gets detected, the vehicle classification system identifies the category of vehicle such as car, bus, truck, etc., then the transaction processing system proceeds the payment procedure. If the transaction completed successfully, then the vehicle was allowed to cross the toll. Also, is there any violation in the toll fee payment, violation enforcement system that detects the number plate of a vehicle.

2.2.2.4.1 Freeway Traffic Management System

The collection of various information using vehicle detection sensors CCTV, the number of vehicles moving around will be calculated and recorded. The tollgate traffic management is getting accelerated with the help of fast cash transfer known as a high pass system which avoids the waiting time at the toll gate [3]. Thus, it helps to minimize the waiting time in toll gates.

2.2.2.5 BUS INFORMATION SYSTEM

It collects the location information of the bus and processes it via the information management system where the centralized information about the bus, its route, arrival time is viewable by the people who are standing in the bus stop which obviously reduces the tension of catching the bus and gives personal satisfaction also [1]. It is a kind of system implemented in Korea.

2.2.2.6 AUTOMATIC FARE COLLECTION

One card is designed to integrate all public transport, such as trains, taxis, buses, and its information [11]. The fare collection is automatically done with the help of that card which makes use of passengers to feel happy and satisfied about the payment.

2.2.2.7 SPEED ALERT SYSTEM

Nowadays road accidents increase due to high vehicle population and over speed. Nearly 1.25 million people die each year due to road traffic accidents [8]; it can be avoided by controlling the speed of a vehicle. Speed alert system gives alert to the passenger and driver when the vehicle speed exceeds the predefined limit. Speed and GPS coordinates of a vehicle calculated frequently, and it can be monitored easily from the traffic control room [7]. If the speed reaches the abnormal limit, then the buzzer alert given to the driver for attaining back to the normal speed. When the speed is not controlled within the specified limit the vehicle information such as speed, GPS position, vehicle number, vehicle color is sent to the traffic control officers through short message service (SMS).

2.2.2.8 TEMPERATURE MONITORING

There is a necessity for monitoring the temperature for the temperature-controlled goods that carry huge loads. The container which carries the loads like meat, foods, and medicines have to be now in a particular temperature. In some cases, the customers are in need of temperature history at the time of delivery of products. This type of control avoids the decaying of goods which is purely dependent on temperature via updating the temperature deviation prior to the industry. EN 12830:1999 logger defines the standard for temperature monitoring.

The temperature monitoring setup consists of two sensors one is for inlet air temperature monitoring and another one is return air. Inlet air is to concentrate on what temperature is to be added to the goods whereas return air concentrates on how much temperature will it sustain in case of rise and fall. The cold transport is well regulated with the alarm system. The minimum and maximum threshold set up in the transported products helps to identify the suspicious temperature deviation and intimates that to the monitoring unit immediately to avoid the damage and penalties. The components that are essential for the tracking of goods temperature is by having temperature sensors, loggers in the cooling unit, on board computers to read the value from the sensor. In addition to the temperature updates to the web office it can be updated directly to the driver via the driver interface application in a mobile phone. Does the sensitive goods are safeguarded with the help of this efficient IoT temperature monitoring system [12].

2.3 ROLE OF IOT IN TRANSPORTATION SECTOR

STS are integrated with the IoT in a stronger way and IoT provides the solution and new direction for the problems which arise in the smart transportation sector. In critical situations, IoT enabled STS to find the optimized and safer routes for travel. A major portion of transport sector revenue is depending on this smart transportation. IoT-adapted STS helps in traffic monitoring, traffic control, smart parking, automatic toll collection, etc. All this application uses a wide variety of sensors, and it paves a way for traffic surveillance and monitoring. There are two main factors that need to be discussed while considering the sensors. They are accuracy and reliability. Accuracy is measured by the number of vehicles counted more than one time as well as the number of vehicles missed in counting and reliability deals with a stable average vehicle count. Sensors are being deployed on the road, in vehicle and on-body sensors on drivers.

2.3.1 ON-ROAD SENSORS

On-Road Sensors is deployed on the road surface and it captures the information regarding traffic flow and helps in travel route prediction. It has high vehicle detection accuracy, and it also uses an effective fault tolerance mechanism [13]. In case any sensor fails while working, to manage that "backup sensors" are deployed on the road. There are various on-road sensors such as Pneumatic Road tube sensor, Inductive loop detector, Magnetic sensor, Piezoelectric sensor, Radar sensor, IR sensors, Ultrasonic sensors, and Road surface condition sensor. The role of road sensors such as: (i) pneumatic road tube sensors are used to track and count the number of vehicles; (ii) inductive loop detector used for absorbing the vehicle presence and its movement; (iii) magnetic sensor used to identify the status of stop and moving vehicles; (iv) piezoelectric sensor is to classify the vehicles and measures the vehicles speed and weight; (v) a RADAR sensor helps to measure the volume and speed of the vehicle along with its motion direction; (vi) IR sensors used for detection of pedestrians in crosswalks; (vii) ultrasonic sensors are for detecting the traffic congestion by sensing the nearby vehicles; and (viii) road surface condition sensor senses the strength of the road construction. Also, on-road sensors have the disadvantage of high sensor installation cost, and traffic interruption during repair, maintenance, installation, etc.

2.3.1.1 PNEUMATIC ROAD TUBE SENSOR

Pneumatic road tube sensor most popularly used in vehicle sensing technology because of its portability it tracks and counts the number of vehicles on the road in an easy way. It is installed below the road surface [14]. In heavy traffic, this helps to sense and count the number of vehicles, speed of a vehicle, vehicle moving direction, gaps between the vehicles, etc. Road tube made up of synthetic polymer material or natural isoprene. Synthetic polymer material is more robust and suitable for all types of environments and natural isoprene is used only for colder climates. Pneumatic road tube sensors operate on the temperature between 15C and 60°C and handle the pressure between 3 and 8 barg. Pneumatic road tubes are used in two forms:

1. **Single Pneumatic Road Tubes:** Counts the number of vehicles crossing the rubber tube as well as the gap between the vehicles.
2. **Combination of Two Pneumatic Road Tubes:** Finds the direction of the traffic and vehicle moving speed.

In Figure 2.1, a single pneumatic road tube is placed on the road surface which gathers the vehicle information.

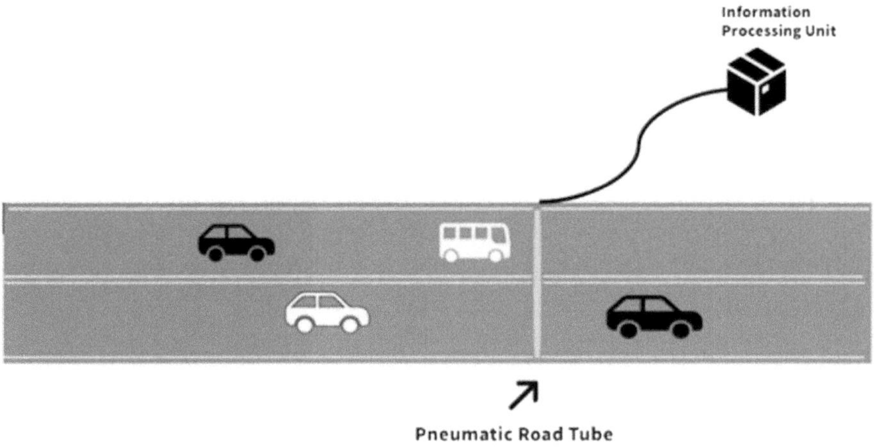

FIGURE 2.1 Single pneumatic road tube.

Air filled rubber tube placed across the road surface, one end of the tube is fully sealed, and the other end of the tube is connected with an Information processing unit. Pneumatic pressure sensors placed inside the rubber tube for sensing the vehicles when a pair of vehicle tiers cross over the tube air pressure inside the rubber tube gets increased. It closes the air switch and generates the electrical energy, and it sends the vehicle related information such as count, speed, vehicle direction, etc., to the Information Processing Unit. Road tubes are sustainable for all sorts of weather conditions and the life span of road tubes varies based on the factor's location, installation, and volume of traffic. Also, it has the problem that a separate road tube is needed for each direction and the collected information needs to be processed or stored in a computer on the roadside. Road tubes are not suitable for higher volume roads and only work well for minimum duration on low volume roads.

2.3.1.2 INDUCTIVE LOOP DETECTOR FOR TRAFFIC MANAGEMENT

Inductive loop detector is an electromagnetic vehicle detection sensor used in traffic management that detects the vehicle which is crossing/arriving at the specified area of installation [15]. Electric current is induced in loop wire by the alternate current or moving magnet which detects the vehicle. Vehicles passing the loop are detected by the loop detector transmitted signals. Loops of wires are placed in the pavement of the road surface and the electronic amplifier is connected with both the ends of the wire. Figures 2.2(a) and 2.2(b) shows the inductive loop detector installed surface and inductive loop detector working, respectively, if the vehicle crosses through the magnetic field of a loop its magnetic field gets affected and this disturbance is sensed by the amplifier and the vehicle gets detected.

The electronic loop detector unit transmits energy into the inductive wire loops at frequencies between 10 kHz and 200 kHz. The inductance gets reduced when the vehicle passes over the loop then the vehicle gets detected. It is also having low manufacturing and transportation costs compared to others. Inductive loop detector size varies by 5 feet to 6 feet, and it takes the installation time between minimum 10 minutes to one hour based on road condition. Most of the conventional loop detectors are used to find the vehicle presence, count the number of vehicles crossing the loop, speed, and gap between vehicles. Nowadays inductive loop detectors come up with advanced technology to classify vehicles. The major drawback of this sensor is at every phase of road maintenance it needs to be reinstalled. Also, traffic disturbance will happen at installation time.

2.3.1.3 MAGNETIC SENSORS FOR DETECTING VEHICLE PRESENCE

Magnetic sensors sense and detect the vehicle presence in the sensor installed surface. It distinguishes the vehicles and allows us to make decisions on traffic light control [16]. Also, it gives the improved solution for heavy traffic based on the sensed data. Magnetic sensors can be placed on the middle or side of the road, when the vehicle or metal object crosses the sensor, it affects the earth's magnetic field, and the metal object or vehicle gets detected. Figure 2.3 depicts the magnetic sensor embedded integrated circuit (IC) which shows the magnetic field distortion by the vehicle. Magnetic sensors are very small in size therefore it has reduced the installation issues and cost.

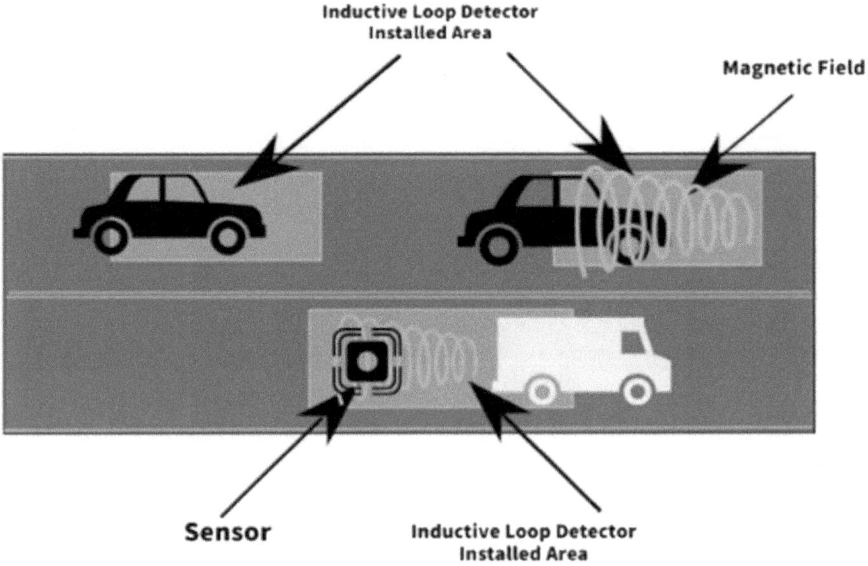

FIGURE 2.2(a) Inductive loop detector installed surface.

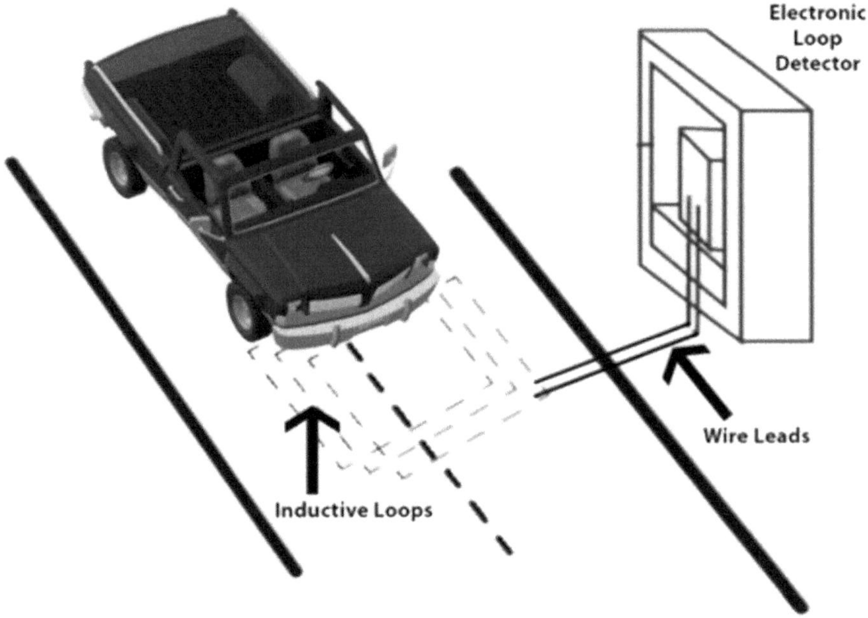

FIGURE 2.2(b) Inductive loop detector working.

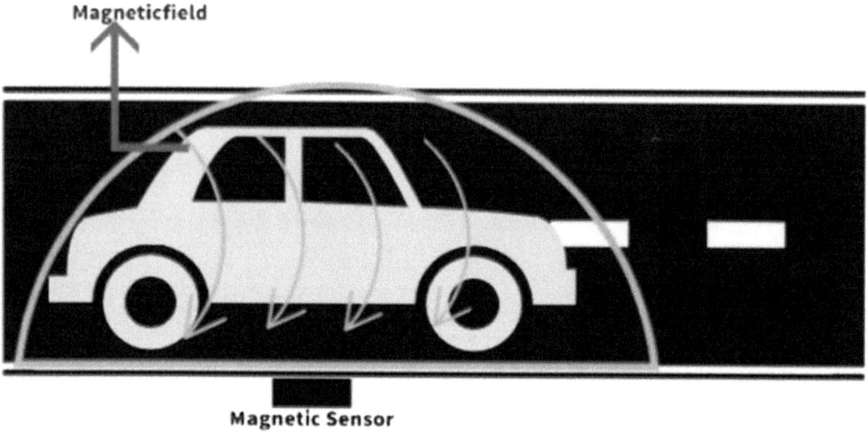

FIGURE 2.3 Magnetic field distortion by the vehicle sensed by magnetic sensor.

2.3.1.4 PIEZOELECTRIC SENSORS TO COUNT THE ROADSIDE VEHICLE COUNT

Piezoelectric Sensors use the piezoelectric effect and identify the changes in pressure, temperature, and force on the sensor environment. Piezoelectric sensors kept under the road surface in a diagonal position [16]. Once the mechanical stress is applied to the sensor piezoelectric effect generates the electricity. When each tier crosses the sensor installed area that changes the voltage pulse. A number of vehicles crossing the sensor is counted by the change in voltage. To count this voltage, a change counting device is connected with the sensor and placed on the roadside. Piezoelectric sensors are made up of quartz material and have high flexibility with small size. This sort of sensor operates on the temperature between 20°C and 60°C and having the sensitivity of 5 V/με. Piezoelectric sensors do not require any external voltage. Once the pressure is applied to it creates a charge and detects the vehicles with the vehicle passing speed of 112 km/h along with weight.

2.3.1.5 RADAR SENSOR FOR MEASURING THE VEHICLE SPEED

Radar sensor represented in Figure 2.4 identifies the objects using the microwave radiation. The transmitted low energy microwave hits the vehicle and is reflected by the object; while reflecting it returns location and speed of a vehicle also it helps in traffic monitoring and protect from vehicle crashes [16]. A RADAR sensor helps to measure the volume and speed of the vehicle along with its motion direction. This sort of sensor is mainly classified into different categories like 24 GHz and 60 GHz based on its range. Object detection technology of 24 GHz and 60 GHz, respectively, detects objects within the range of 100 m and 10 m. Autonomous cars speed is controlled by this radar sensor; in the varying environmental conditions, speed gets slowed down also the sensor will work effectively. Radar sensors will not be affected by the weather conditions like rainy, windy, foggy, etc. It is more suitable for outdoor applications also due its withstanding physical surface hardness.

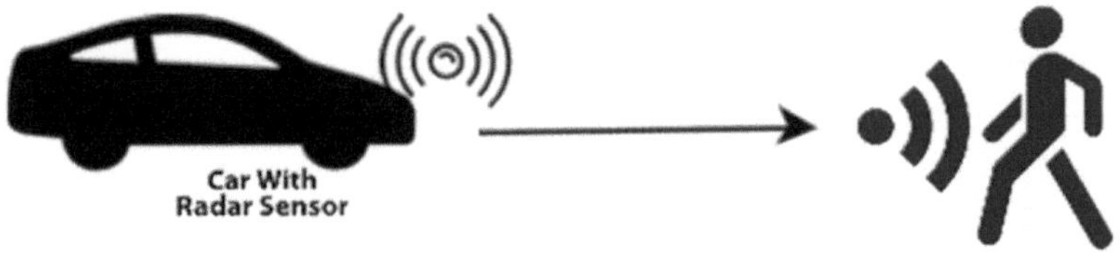

FIGURE 2.4 Radar sensor object detection.

2.3.1.6 INFRARED SENSOR (IR) FOR PEDESTRIAN DETECTION

Infrared sensor (IR) detects the people's presence using infrared radiation in the surrounding environment. It is simply called an optical sensor. IR sensor used for the detection of pedestrians in crosswalks [16]. The light beam gets interrupted and breaks the continuity of it if any of the objects crosses over it. It prevents the unknown hits on the people who are walking on the roadside. The IR sensor detects the object as shown in Figure 2.5 with the maximum sensing range of 200 meters. The IR sensor requires the mounting structure for both transmitter and receiver end for the stability and constant working without any disturbance.

2.3.1.7 ULTRASONIC SENSOR FOR ESTIMATING TRAFFIC FLOW

Ultrasonic sensors are for detecting the traffic congestion by sensing the nearby vehicles; it measures the distance between it and other objects which are on its line of sound [16]. With the echo from the object, the distance is estimated as shown in Figure 2.6. The maximum sensing range of the ultrasonic sensor is 4 meters. The road surface is considered as a fixed and stationary object whereas the vehicles are considered as the moving object. The sensor is employed 4 m from the road surface and positioned in mid periodically along the vertical alignment. If any of the vehicles crosses this area, then the distance between the sensor and the road surface is getting reduced, thus ensuring the vehicle presence and its occupancy period to estimate the traffic flow. Here the length of the vehicle is fed as an input accordingly to calculate the speed of the vehicles moved along the line of sound of an ultrasonic sensor and the count of it is acquired for an intelligent transportation system. The traffic information is estimated in this way using an ultrasonic sensor.

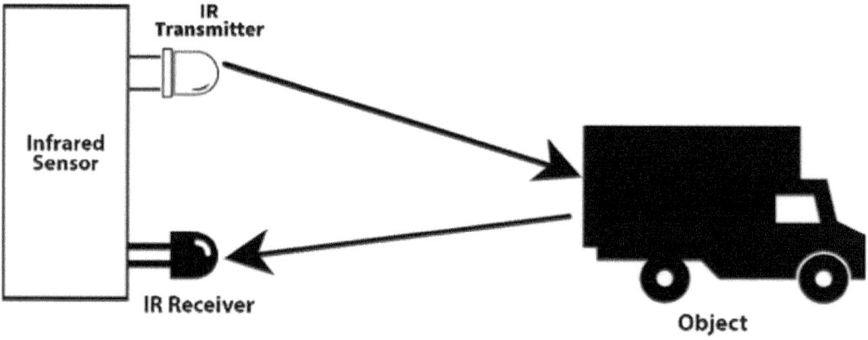

FIGURE 2.5 Infrared sensor object detection.

The sensors mounted on both the sides of the roads will enhance the efficiency of detecting the vehicle over a single side mounting device. The sensor network configuration is quite challenging for the poor road infrastructure. The horizontal placement of sensors will register the non-vehicles in a vehicle category and provide the data in an inappropriate manner. It happens at times and can be eradicated by placing the sensors on the edge of the roads. The road surface has to be smooth to avoid the case of absorption and dispersion of sound signal emitted from the ultrasonic sensor. For accurate traffic flow estimation and avoid congestions, the sensors are overhead mounted for covering wider road networks. HCSR04 is the one which is applicable to measure the object which is at a distance from 2 cm to 400 cm. The ultrasonic sensors can be mounted on the parking slots for detecting the vehicles to ensure the slot freeness; thus it is possible for efficient parking management also. The ultrasonic sensor is used to detect the disorder in the road based on the decayed reflection intensity of ultrasonic waves. It produces alarm sounds in case of dangerous road conditions. It saves more lives of the human with its easy configuration and availability [17].

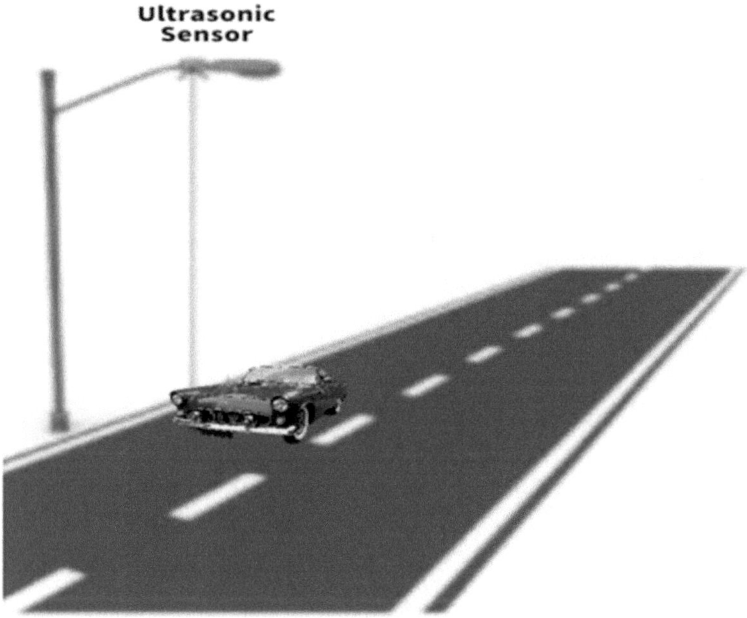

FIGURE 2.6 Ultrasonic sensor for vehicle detection.

2.3.1.8 ROAD SURFACE CONDITION SENSOR

Road surface condition sensor senses the weather condition and strength of the road construction [16]. The abnormalities of the road while traveling is not observed all the time by the drivers, especially during night times. This leads to many accidents and complications at the driver's end. The road surface condition sensor is used to sense and assess the pavement surfaces, roads, highways, bridges, and parking areas. It also supports the elderly people and the visually impaired people to know the road surface by prior alertness.

There are three configurations.

- The sensor is mounted on the pavement surface with the drilled cavity even though working to face the huge traffic scenario in real-time;
- Non-invasive pole mounted installation to measure the road surface; and
- The real time surface condition sensing is with remote sensing via the sensor mounted on the vehicle.

Puddle, asphalt, and lawn or soil are the types of road condition. The acceptable form of road condition is asphalt because the chance for getting slip and fall is more possible for the aged and visually affected people in puddles. This way of distinction is very helpful and supportive for the physically handicapped people who are coming in wheelchairs also.

AI enabled features of all the above road sensors will observe the road condition and accurately collect data that is driven and analyzed for ground transportation in a better way.

2.3.2 IN-VEHICLE SENSORS

In vehicle sensors are used to observe the surrounding environment while driving it helps for the driver to ensure safety. Various in-vehicle sensors are shown in Figure 2.7. The sophistication of IoT is estimated with the help of sensors and its integration unit.

2.3.2.1 NIGHT VISION SENSOR

Major accidents happen unknowingly because of the less vision at the driver's end. Human vision is quite poor at nighttime when compared to the animal's vision. It is made possible with the help of appropriate electromagnetic spectrum and light intensity. To see the person's image even at a distance the thermographic cameras are used [19]. It is an additional option embedded in the advanced vehicles. The IR and thermal sensor are mounted in the vehicle and controlled with the onboard computers employed on the center console. The object detection and its distance are sensed with the help of IR and displayed in the display device which helps to circumvent the obstacle in front of the vehicle sooner before reaching it. The sensors can be placed anywhere in the vehicle like windshield, headlights, etc.

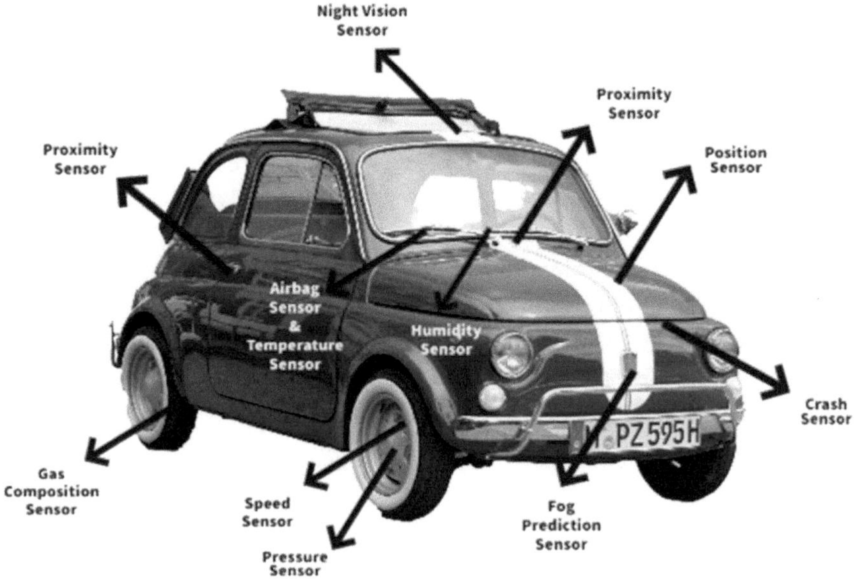

FIGURE 2.7 Sensors placed inside car.

There are two categories involved in night vision sensors. They are:

1. **Active Night Vision Sensor:** It is a sensor supported to know about the objects which are viewable around 600 feet distance with the help of IR sensors. One of the quiet drawbacks is that the weather condition affects the sensor's capability of visualizing.
2. **Passive Night Vision Sensor:** In passive night vision sensors, there will not be any clear image of Infront objects but still supports for the extended range of 1,000 feet. It purely depends on the thermographic camera, and it is unreliable if the temperature is high.

2.3.2.2 SPEED SENSOR

It is the sensor which is used to read the speed of the vehicle using wheel rotation. It is usually an optical sensor in which the light signal along with its reflection is considered [20]; Inductive sensor in which the magnetic pulse is getting generated and is in the form of a wave;

The efficiency of the operating functionalities of the vehicles is reflected by using speed sensors for Automotive vehicles.

There are various aspects to consider to be more useful. Some of them are:

- The sensor input helps to control the pressures on the power steering to slow down the speed in case of parking areas.
- The anti-lock braking system which is simply called ABS utilizes the speed sensor data and helps to identify when the wheel gets locked up and releases a certain amount of pressure to maintain the stability with respect to the direction of the vehicle it is moving.
- The electronic engine control systems also will make use of vehicle speed sensor data.

As a whole, it helps to find the fault which occurs due to uncontrollable speed via proper signaling from a speed sensor. It is deployed in the wheel to measure and compare the wheel rotations and take necessary actions further with the sensed data. The data is given to the ECU unit to instruct the exact timing of when the gear has to be changed. Since the automatic transmission shift sensor and speed sensor are having the correlated functionalities, the fault of the vehicle is also easily determined.

2.3.2.3 PROXIMITY SENSOR

The nearby and closest object is detected with the help of a proximity sensor. The sensed measurements are highly useful for Automatic Cruise Control. The speed of the vehicle is purely based on the vehicles in front and behind on it [20]. It is very much useful for the driverless cars to take necessary actions with the detected signal. There are different types of proximity sensors. They are inductive, capacitive, and photoelectric. (i) The electromagnetic field generated from an inductive type of proximity sensor is changed, if the metallic target is entering into the field which reflects a varying proportion of the field. (ii) The capacitive type is not only for metal targets but also for other forms like glass, liquid, etc. (iii) Through the breaks in the beam of light between sender and receiver the targets and their presence are detected with the help of photoelectric proximity sensor. The most common forms are magnetic and ultrasonic proximity sensors.

2.3.2.4 AIR BAG

The airbag sensor is a passive protection device where the human interruption is not required and designed only for the cars, not for the motorcycles. The signals from the air bag sensor which is a MEMS accelerometer rely on the control unit which enables the safety features when at risk [21]. There are two common forms of air bag sensor one is electrical which is purely a battery operated and another one is mechanical which works independent of electrical energy source. The generic ball and tube mechanism is an electromechanical one which depicts the automotive reaction while facing the collision. The tube consists of a small electric circuit switch and the ball/roller type weighted coil is attached together with a magnet. If the accident occurs, then the ball is dispatched and gets rolled in the forward direction which hits the switch and makes the circuit to be closed to activate seat belt locks and automatic door locks, etc. At the same time, the inflator system gets the signal and starts to fill up the nitrogen gas through chemical charging and explode it as an aid. The impact of the sensed data is analyzed with the help of AI crash algorithms. The crash algorithm designed for the electronic control unit of Airbags prevents the vehicle occupants by enabling the number of airbags to be busted at that point of time. The readiness and the response time is just a few milliseconds.

2.3.2.5 PRESSURE SENSOR

The pressure sensor is used to sense useful vehicle fluids like gearbox and engine oil, braking system's hydraulic oil, oil in cooling unit and fuel system oil. It is based on a piezo resistive effect in which the change in resistance causes deflection due to the fluid pressure. The targeting features make the sensor acceptable to do the designed task only when withstanding the high temperature, pressure, dust air environments at the extreme end [22]. The sensing ability of a pressure sensor is from 0 bar to 600 bar with the operating temperature of –40°C to +125°C. Recent inventions on pressure sensors are highly useful to circumvent from the situations of pedestrian hits to be normal by giving the forceful airy barrier if they hit in the front face of the vehicle. It is by sensing the deformity of the bumper as soon as the pedestrian slides on it and enables the safety system of the car via activating the reserved compressed air in the bay.

2.3.2.6 GAS COMPOSITION SENSOR

The air quality is observed with the help of gas composition sensors [23]. The gasses like carbon monoxide (CO) and oxygen are sensed with the help of gas sensors. If in case the dangerous gas concentration is exhausted the limit, then it will be alerted to the driver via the alarm sound. The uniformity of ventilation and air conditioning is maintained by continuous monitoring of the gas present in the circulating air of the vehicle surrounding the environment. The sensor should leak proof and water resistant to operate from – 40°C to 85°C. It tends to support vehicular fuel efficiency.

2.3.2.7 TEMPERATURE SENSOR

Temperature sensors are in need for the cold transport systems such as food transportation using the container milk, meat, and medicines; all need to be maintained in a particular temperature to prevent the items from decaying. The in-car temperature sensor is a negative temperature coefficient sensor in which the resistance value decreases when the temperature increases [18]. The signal voltage acceptable for this temperature sensor is 0 volt and 5 volts. It will sense the vehicle into two verticals one is in the passenger compartment and another one is air temperature outside of the vehicle. The temperature sensor is acting as an input for many units available in the car such as fan speed, cooling loads, Sun loads which are adjustable based on the varying temperature. LM35 is the temperature sensor used here which has three leads (power supply, ground, and output voltage) to measure the ambient temperature. ESP 8266 Wi-Fi module used to feed the temperature sensor's data to the web. The battery lifetime of the sensor plays a vital role for the sensor working. There is a wireless battery free sensor called RFM 3200 which has a longer lifetime compared to others. The advantages are the complete eradication of nonlinear temperature problems through which it ensures high accuracy.

2.3.2.8 POSITION SENSOR

The safety requirements for high end vehicles are ensured by detecting the abnormalities in a rapid fashion through the sensor technology. The power steering electric cars are in need of calculating the position to decrease the fuel consumption and CO_2 emissions. To detect the angular rotation of the steering wheel and control the motor, the position sensor is deployed in the vehicle [24]. The precise controlling of torque both at starting and high-speed operations is possible by fixing the position sensors in electrically commutated motors.

There are three types of position sensors. They are hall effect sensors, optical sensors, and inductive position sensors:

1. **Hall Effect Sensor:** It is used to sense the position of the pedal. It detects the effect of resistance on the current carrying conductor with respect to the magnetic field applied. The IC-based AS5262 is a Hall-based position sensor that produces 360° angular measurements accurately. Its operating temperature is – 40°C to 150°C. It is more immune to unnecessary stray fields via proper surface magnetization.
2. **Optical Sensor:** It consists of phototransistors to sense the movement of optical marks. It is a crankshaft position sensor for sensing the light being reflected from the shaft's surface which is in rotation. The change can be sensed with the light and dark marks of the optical beam.
3. **Inductive Pedal Position Sensor:** The digital component is used to drive the coil with high current. The consecutive creation of the magnetic field and its relative current phase is used to determine the position of the rotating coil using the accelerator pedal position sensor.

2.3.2.9 HUMIDITY SENSOR

It is an electronic device known as a sensor which senses the surrounding environment's air humidity which could form mist [25]. If the sensed amount of water vapor present in the air (mist) quantity on the glass of the vehicle is considered to be more then it will be difficult for the people to see the objects which are in front of the vehicles. The IoT intelligence is proven by integrating one sensor's data with the other sensor to react and thus leads to optimization. The air humidity sensor data is given to the temperature sensor to measure the temperature at which the concentration of humidity is high. So, with the help of a climate control unit the dew point at which temperature the mist becomes high is calculated and regulates it with a condenser of the air conditioning unit present inside the vehicle. The major advantage of this battery enabled sensor is to make the people inside the vehicles drive in a comfortable way through its features like durability of 15 years, small size, reliability, etc. It further leads to avoiding accidents in the hilly areas majorly.

2.3.2.10 FOG PREDICTION SENSOR

The self-driving vehicles and as well as the human driving vehicles are facing difficulty in the poor weather condition like morning fog. The fog sensor senses the level of moisture content lying on the glass surface inside the vehicle [26]. The fog sensor which is in small size will generate the infrared light on its own to detect the presence of fog. The detection is done with the help of the scattering phenomenon of the light beam. The scattered beam of emitted IR is again received by a fog sensor if the fog is formed in the interior glass. The observed data from the sensor is fed to the climate control unit present inside the vehicle and defog it with the help of any heating effect. The fog sensor is highly accurate and detect it before it is being visualized by the human.

2.3.2.11 VEHICLE CRASH SENSOR

There are many blind spots which are not viewable for the driver if they are in the driver's seat. There are various sensors which are useful to detect the obstacle before an accident occurs [27]. Electromagnetic sensors, Ultrasonic sensors are the few sensors useful to detect the obstacle and alert the driver when the vehicle is on the move.

- Using an electromagnetic sensor employed in the bumper will emit the signal to indicate the obstacle which is nearby to the vehicle. If the vehicle is continually maintained at the same speed without noticing the emitted signal indication of the sensor, then the sound signal is produced to alert the driver. This process is considered to be a smooth and slow process.
- Using the wireless ultrasonic sensor, the distance of the obstacle is detected by the echo sound of the generated radio signal from the obstacle. It can be placed in the front and as well as back parts of the vehicle.

- Reverse car parking sensors will be activated if the car is in reverse gear mode. The sensor will generate ultrasonic radio waves to transmit and receive the reflected signals from the obstacle.

2.3.3 ON BODY SENSORS

The drivers are spending most of their lifetime in the vehicles and their driving manner brings the happiness for many people for reaching the destination on time. Still the continuous workloads and their competency to drive the vehicles on a timely basis makes them run continuously and induces some sort of aggressiveness. Alcohol consumption, injected drugs, drowsiness, inattention driving, speediness are the major causes for the vehicle crash and people death due to those accidents. Due to this the traffic safety is also affected and put in danger. It has to be taken care of by continuously observing the behavioral actions of the driver and ensures whether the driver is driving the vehicle with proper control and stability [28]. The sensed physiological signs help to avoid accidents happening due to the inadequate sensible actions of the driver.

The emotional disorders are determined with the help of clues which have been programmed in a sensing unit. The role of on body sensors such as EOG, ECG, EEG, EMG senses the abnormalities in driver's health conditions and helps to ensure safe traveling. On-body sensors are placed in various parts of the human body are shown in Figure 2.8 and it detects driver distraction, drowsiness, emotions using monitoring units. The frequency of yawning, eye blinking, head movement is predicted with the help of a combination of sensors. The EOG (electrooculogram) sensor is used to detect the eyeball movement; It detects the eye blinks up to 10 cm range. EMG is used to detect the skin's electric potential; EEG (electroencephalogram) sensor is for observing the brain activity of the driver which consists of electrodes to analyze the alpha, beta, and gamma waves induced in the brain stimuli to detect the drowsiness and fatigue of the driver. This helps to provide safety to both drivers and passengers. The heart rate of the driver is picked up for periodic intervals of time by using an ECG sensor. It consists of two electrodes employed on the body which is in contact with the driver's skin directly but not invasive. There is an option to make it non-contact by positioning it in the driver's seat back also. Because it is possible to get the ECG and EEG signal even with their hair or cloth without the skin contact [28].

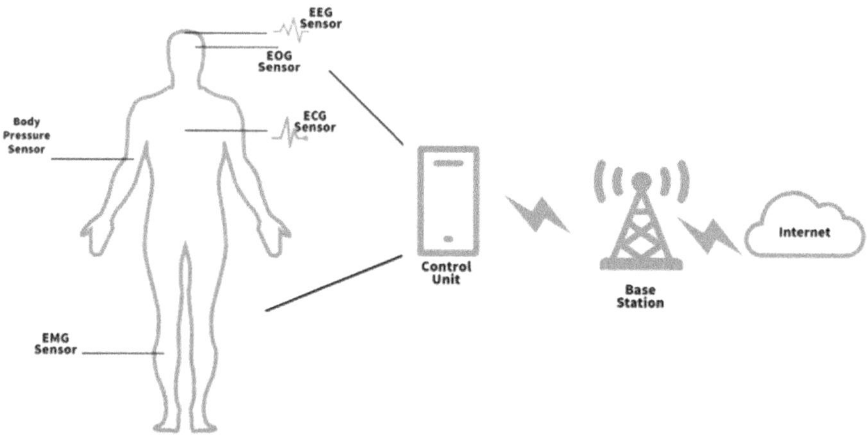

FIGURE 2.8 On body sensors.

The variance in electrical activity and irregularity in the sensed heartbeat signal shows that the driver is quite inactive through drowsiness. The signal acquisition module and the signal conditioning unit which holds the filter circuit to remove the unwanted signal and amplifier circuit to strengthen the signal will help to reach the signal processing unit through wireless communication. This further leads to making the driver fall asleep using proper alertness.

Alcohol detection – using the PAS 328 alcohol sensor the amount of alcohol consumption is detected using the breathing. This setup is connected to the fuel supply unit which cuts down the fuel supply to stop the vehicle. If the 10% of alcohol is detected, then the vehicle will automatically be directed to the speed of 80 km/hr. If it is between 20% and 30% the speed of the vehicle is maintained to be 60 km/hr. If it exceeds 30%, then the alert message is triggered and displayed in the controller unit LCD screen. In case the message is ignored and does not stop the vehicle within a few minutes then it directly passes the short message along with the location of the driver using GPS to the regulatory commission which takes care of this.

2.4 ROLE OF AI IN TRANSPORTATION

One such way to incorporate human intelligence in machines is known to be AI. The conceptual contribution of AI impacts the transport sector widely. The AI technology deliberately contributes to passenger safety, reducing traffic congestion and accidents. The important considerations of AI are safety, efficiency, reliability, and pollution. To make the intelligence of transportation a bit higher it will be embedded to have autonomous functionalities in buses, trains, cars, and marine vehicles. The fuel consumption is possible with the help of indication about the road traffic status in prior while opting the route to reach destination. Advancements of it are proven while dispatching the loads carried by consecutive trucks are being in control of its accelerator and brake through AI is made possible even in case of complex traffic. The integration of IoT and AI administering the recent marketing trends to clear the traffic and maintaining the actions of vehicles in a smooth way.

In transportation AI technology it is mainly involved in self-driving vehicles, traffic light control and pedestrians path identification [29]. In urban areas, it can be used to predict the decision from the previously learned data and accidents can be avoided in a major way. In critical situations when the decision is unpredictable, AI methods like neural networks, genetic algorithms (GAs) can be used for prediction. In transportation, optimized routes are identified priorly. It reduces the transportation cost and provides safety.

1. **AI in Public Transport:** In countries with high populations, public transportation becomes the greatest gift. In AI, machines act with human intelligence [29]. Deployment and need of AI are classified under different categories such as learning, prediction, and automation. In the learning stage opinion/feedback is collected from the public environment. During prediction, the transportation model identifies the congested route and provides the safe alternate path. Once the learning and prediction gets over, the entire process is automated for solving the forthcoming problems.
2. **Autonomous Aircraft:** In autonomous airplanes at the time of boarding, to verify the identities of the passengers AI uses the facial recognition technology. Scanning processes are automated and the passengers can scan their face and it gets verified automatically without any human intervention. Also, at the time of climbing, airplanes will use the highest rate of fuel. To measure the fuel consumption level, a pilot climb profile is created along with flight information. AI models analyze the pilot's profile and predict the expected fuel consumption priorly.

3. **Goods Delivery:** To avoid the issues in goods transportation and delivery, drones with AI technology are deployed; it avoids the human error. Drones used for delivering the product in customer's doorstep. Image processing and Neural Network technology used for detecting the objects/obstacles such as hills, buildings, etc., in the flying environment. Sensors and cameras are used for capturing images. Once the images get captured, it is processed, and some useful decisions are taken from this processed information.
4. **Traffic Management:** In traffic management, AI creates the environment friendly methods for smooth transportation. Traffic lights are embedded with AI technology based on traffic congestion and it will take decisions automatically [29]. AI recognizes the environment and acts accordingly for taking decisions.

2.4.1 IMPLEMENTATION OF INTEGRATED IOT AND AI TECHNIQUE FOR DRIVER MONITORING

The safety-enabling technology is successfully implemented with a test model in which the IoT-enabled sensor Network is configured and its output data is analyzed with the help of AI technique. The eye blink sensor consists of an IR transmitter and receiver detects the number of eye blinks using the reflected rays from the eyes. The sensed data from the eye blink sensor acts as a test data and is driven to the trained model to identify whether the driver is in a drowsy state or not. If the eye position of the driver remains closed for a specific period due to drowsiness, then it may be a cause for the accidents; this can be prevented by alerting the driver then and there in case of any lethargic driving.

In Figures 2.9(a) and 2.9(b) in which the eyes are being opened and closed for some time. Figure 2.9(b) illustrates the intentional eye closure and indicates the drowsy feel direct to less concentration.

Using this model, the region of interest (RoI) is created with the help of detected faces which are then fed to the CNN classifier. It classifies the eyes based on open and closed condition with its obtained values. The normal person blinks the eye 10 times in a minute. According to this, the threshold is fixed based on the time which is 15 seconds. If the obtained value is exceeded the threshold value, then it confirms the driver's state as drowsy and the alert sign will be indicated through an alarm signal and display notification to the driver.

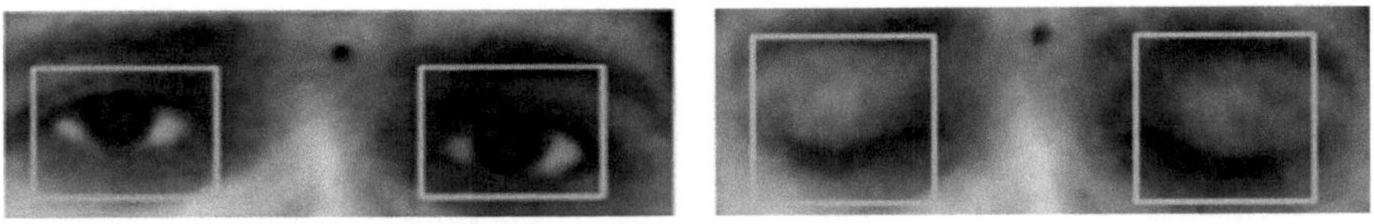

FIGURE 2.9 (a) Driver's opened eyes; and (b) driver's closed eyes.

The model provides the accurate detection and classification even in midnight condition. Thus, the model efficacy saves the passengers and ensures safe driving all the time.

2.5 CONCLUSION

The revolution of IoT and AI in the Intelligent Transportation System is reviewed with several IoT sensors and AI techniques. And structured the way of managing the transportation in three verticals as on road, in vehicle and on body of driver. The detailed role and significance of AI and IoT in ITS are highlighted and implemented as a model. It has CNN classifier which shows its readiness to classify the driver's drowsiness feels quite an advantage for the sensor enabled network using IoT. Thus, the implemented model, its detection efficiency and the estimation time proves their intelligence for the ITS. Likewise, the probing of each and every thing in transportation leads to a better management of it autonomously.

KEYWORDS

- **artificial intelligence**
- **automated license plate recognition**
- **infrared sensor**
- **integrated circuit**
- **Internet of things**
- **short message service**

REFERENCES

1. Boshita, T., Suzuki, H., & Matsumoto, Y., (2018). IoT-based bus location system using LoRaWAN. In: *2018 21st International Conference on Intelligent Transportation Systems (ITSC)*, pp. 933–938. doi: 10.1109/ITSC.2018.8569920.
2. Zhu, F., Lv, Y., Chen, Y., Wang, X., Xiong, G., & Wang, F. Y., (2020). Parallel transportation systems: Toward IoT-enabled smart urban traffic control and management. In: *IEEE Transactions on Intelligent Transportation Systems* (Vol. 21, No. 10, pp. 4063–4071). doi: 10.1109/TITS.2019.2934991.
3. Ahmed, S., Tan, T. M., Mondol, A. M., Alam, Z., Nawal, N., & Uddin, J., (2019). Automated toll collection system based on RFID sensor. In: *2019 International Carnahan Conference on Security Technology (ICCST)* (pp. 1–3). doi: 10.1109/CCST.2019.8888429.
4. Tran, T. K. O., Tran, N. H., Pham, C., LeAnh, T., Thai, M. T., & Hong, C. S., (2020). Parking assignment: Minimizing parking expenses and balancing parking demand among multiple parking lots. In: *IEEE Transactions on Automation Science and Engineering* (Vol. 17, No. 3, pp. 1320–1331). doi: 10.1109/TASE.2019.2948200.
5. Mohammed, S., Supriya, P., & Medhat, A., (2020). Development of an IoT based real-time traffic monitoring system for city governance. *Global Transitions, 2*, 230–245. ISSN 2589-7918, https://doi.org/10.1016/j.glt.2020.09.004.
6. Smith, A., (2008). *Emergency Vehicle Wireless Routing and Notification System.* SAE Technical Paper 2008-36-0079. https://doi.org/10.4271/2008-36-0079.
7. Kodali, R., & Sairam, M., (2016). *Over Speed Monitoring System,* 752–757. 10.1109/IC3I.2016.7918061.
8. https://www.afro.who.int/health-topics/road-safety (accessed on 1 January 2023).
9. Ahmad, I. S., Boufama, B., Habashi, P., Anderson, W., & Elamsy, T., (2015). Automatic license plate recognition: A comparative study. In: *2015 IEEE International Symposium on Signal Processing and Information Technology (ISSPIT)* (pp. 635–640). doi: 10.1109/ISSPIT.2015.7394415.
10. Abduljabbar, R., Dia, H., Liyanage, S., & Bagloee, S., (2019). Applications of artificial intelligence in transport: An overview. *Sustainability, 11*, 189. 10.3390/su11010189.
11. Ampelas, A., (2001). Automatic fare collection. In: *2001 IEEE Intelligent Transportation Systems: Proceedings (Cat. No.01TH8585)* (pp. 1164–1166). doi: 10.1109/ITSC.2001.948827.
12. Rajesh, K. D., & Manjup, P. S., (2013). Cloud based M-healthcare emergency using SPOC. In: *2013 Fifth International Conference on Advanced Computing (ICoAC).*
13. Chan, K. Y., & Dillon, T. S., (2013). On-road sensor configuration design for traffic flow prediction using fuzzy neural networks and Taguchi method. In: *IEEE Transactions on Instrumentation and Measurement* (Vol. 62, No. 1, pp. 50–59). doi: 10.1109/TIM.2012.2212506.
14. Soumya, R. J., Raju, S., Rajesh, K. D., & Kavita, S., (2019). Recent advances and future research directions in edge cloud framework. *International Journal of Engineering and Advanced Technology, 9*(2), 439–444.
15. Mills, M. K., (1981). Inductive loop detector analysis. In: *31st IEEE Vehicular Technology Conference* (pp. 401–411). doi: 10.1109/VTC.1981.1622962.

16. Krishnamoorthi, S., Jayapaul, P., Dhanaraj, R. K., et al., (2021). Design of pseudo-random number generator from turbulence padded chaotic map. *Nonlinear Dyn.*
17. Obed, A., Ebenezer, Q., & Eric, O., (2020). Ultrasonic sensor-based traffic information acquisition system; a cheaper alternative for ITS application in developing countries. *Scientific African, 9*, e00487, ISSN 2468-2276. https://doi.org/10.1016/j.sciaf.2020.e00487.
18. https://www.freeasestudyguides.com/ac-in-car-temperature-sensor.html (accessed on 1 January 2023).
19. https://electronics.howstuffworks.com/gadgets/high-tech-gadgets/nightvision.htm (accessed on 1 January 2023).
20. Guerrero-Ibáñez, J., Zeadally, S., & Contreras-Castillo, J., (2018). Sensor technologies for intelligent transportation systems. *Sensors. 18*(4), 1212. https://doi.org/10.3390/s18041212.
21. https://trid.trb.org/view/719940 (accessed on 1 January 2023).
22. https://www.mdpi.com/1424-8220/20/2/451/htm (accessed on 1 January 2023).
23. Arjen, B., Jörgen, S., & Huib, B., (2016). Gas composition sensor for natural gas and biogas. *Procedia Engineering, 168*, 197–200. ISSN 1877-7058. https://doi.org/10.1016/j.proeng.2016.11.216.
24. https://en.wikipedia.org/wiki/Position_sensor (accessed on 1 January 2023).
25. Hamid, F., Rahman, W., & Mohd, N. H., (2014). Humidity sensors principle, mechanism, and fabrication technologies: A comprehensive review. *Sensors, 14*, 7881–7939. doi: 10.3390/s140507881.
26. Sallis, P., Dannheim, C., Icking, C., & Maeder, M., (2014). Air pollution and fog detection through vehicular sensors. In: *2014 8th Asia Modelling Symposium* (pp. 181–186). doi: 10.1109/AMS.2014.43.
27. Ramasamy, M. D., Periasamy, K., Krishnasamy, L., Dhanaraj, R. K., Kadry, S., & Nam, Y. (2021). Multi-disease classification model using Strassen's half of threshold (shot) training algorithm in healthcare sector. In: *IEEE Access*.
28. Murugan, S., Selvaraj, J., & Sahayadhas, A., (2020). Detection and analysis: Driver state with electrocardiogram (ECG). *Physical and Engineering Sciences in Medicine, 43*. 10.1007/s13246-020-00853-8.
29. https://www.lanner-america.com/blog/examples-artificial-intelligence-applications-transportation/ (accessed on 1 January 2023).

CHAPTER 3

Smart Air Pollution Monitoring and Prediction System Using Deep Learning Enabled IoT Technology

K. M. KARTHICK RAGHUNATH[1], K. S. ARVIND[1], K. S. SUGANYA[2], and R. VINOTHSARAVANAN[3]

[1]*Associate Professor, Department of Computer Science and Engineering, Jain University, Bangalore, Karnataka, India*

[2]*Assistant Professor, Department of Information and Technology, Bannari Amman Institute of Technology, Sathyamangalam, Tamil Nadu, India*

[3]*Assistant Professor, Department of Computer Science and Engineering, JAIN University, Bangalore, Karnataka, India*

ABSTRACT

The series of issues with the engines of modern-day automobiles is growing at an alarming rate. Problems emerge when emissions exceed acceptable limits from the cluster of vehicles in a particular region. Despite the fact that fuel usage has changed significantly, growing metropolises and industrialized scenarios are still considered significant factors that cause air pollutants. DL advances have made it feasible to create prediction systems relying on facts (real-time data) for pollutant evaluation and control. In this chapter, vehicle emissions are being tracked through the internet of things (IoT) technology. The real-time data is used to create a prediction

Artificial Intelligence for Future Intelligent Transportation: Smarter and Greener Infrastructure Design. Rajesh Kumar Dhanaraj, Nilayam Kumar Kamila, Subhendu Kumar Pani, Balamurugan Balusamy, & Vani Rajasekar (Eds.)

model that forecasts carbon monoxide (CO) levels in a particular region of interest (RoI). Thus, it aids in diversion strategies of traffic flow from polluted to normal regions. For this, vehicles passing through the RoI are equipped with sensors that monitor pollution levels. Traffic regulators are informed of the prevailing pollutant condition in the concerned RoI, and the impact of the vehicles are monitored by utilizing ensemble learning (EL) methods. EL utilizes several base classifiers to refine the accuracy level in prediction. When predicting the pollution load at a particular RoI, an EL model uses historical data together with the most recent data from sensors to make its predictions. Thus, a smart emission monitoring system may be used to keep an eye on the pollution level while keeping it under control and support managing the traffic flow and pollution load at the concerned RoI.

3.1 INTRODUCTION

Air pollution is a big issue in modern civilization, especially as the number of people moving to cities grows. Consequently, considerable efforts are focused on developing viable monitoring systems, which include sensors and cameras that collect data in real-time [1]. Air pollution has been found to have a significant impact on human health in industrially developing or developed nations when there is no infrastructure in place to monitor or regulate it. According to recent studies, there is a strong link between air pollution and ailments such as asthma and lung disease. The World Health Organization (WHO) has established rules to limit the levels of specific gases such as nitrogen dioxide (NO_2), sulfur dioxide (SO_2), and ozone (O_3) in the atmosphere. Air quality monitoring (AQM) stations, which are very exact and accurate, are used to calculate the air quality index (AQI) and monitor pollution. They are extremely accurate and reliable in monitoring a wide spectrum of air contaminants.

However, these stations still trail in three areas: (i) infrastructure, which is required for installation because of the large size; (ii) operational needs are rather sophisticated; and (iii) ongoing costs of setup, daily maintenance, and calibration. Due to these faults, the number of installations is reduced, resulting in discrete AQM Networks with little data on air pollution. The use of wireless network technologies in monitoring diverse real-time sensor data, such as temperature, pressure, and air pollution, has been allowed by recent improvements in embedded electronics. The devices' data is wirelessly transferred to a database on a distant server for further analysis and processing. Because of wireless data transmission, the notion of using small, inexpensive AQM motes has surely grown in popularity. The purpose of this study is to discover a good forecast and predict a prototype or model for the decrease or increase of certain air pollutants such as CO, O_3, SO_2, and NO_2, all of which are deemed detrimental by WHO recommendations. For time-series forecasting, three machine learning (ML) methods were utilized and tested: Linear Regression, Random Forest, XGBoost, and ARIMA. All of these models have a wide variety of applications and have high predictive ability as well as generalization power.

By the year 2050, a minimum of 70% of the world's population will live in cities, necessitating efficient systems for monitoring and forecasting air pollution. The least contaminated regions are North America and Australia, followed by Central Europe, while the most polluted are India and Asia. Even though there isn't a lot of data for Africa and South America, data that is available suggests that pollution levels of air are also high in that region. According to the WHO research, every year, 6,00,000 children under the age of 15 die as a result of air pollution. According to studies, air pollution kills 6.9 to 7.1 million people each year, and 91% of the population of the globe breaths toxic air. Children under the age of five in underdeveloped countries breathe filthy air at a rate of 98%. The global economy lost $5 trillion in welfare due to premature mortality caused by air pollution.

The internet of things (IoT) paradigm has emerged in the last few years, digital communication transceivers, microcontrollers, and an appropriate protocol stack that lets them to connect with one another and with users, allowing them to become a vital component of the Internet. Devices like household appliances, surveillance cameras, sensors, and cars may now create massive volumes of data that can be evaluated and used to develop new applications thanks to this feature. Recent advancements in the field of IoT technologies have enabled people to create effective methods that consist of many sensors connected wirelessly to monitor various air contaminants in different locations, either through Wi-Fi or Bluetooth for adjacent mobile networks or regions, and satellites for remote or isolated regions. To assess developing trends and enable near real-time monitoring, these amounts of different data may require parallelization employing big data and cloud computing technologies [2]. Monitoring systems enable us to obtain highly qualified data that may subsequently be utilized to gain a better understanding of pollution.

It is expected that air pollution forecasting systems would use the AQI, which describes each pollutant's value, such as particle matter (PM) or carbon dioxide concentration (i.e., CO_2, PM10, PM2.5, etc.), to identify sites with a lot of pollution. Governments may use predictive air pollution technology to help them develop more intelligent solutions and preventative measures to address air quality challenges. Although the literature contains a variety of models and methodologies for forecasting air contamination, they are in general be classified by two main groups. Firstly, models or prototypes for tracking pollution production, dispersion of pollution, and transmission. These models' prediction outcomes are generated by numerical simulations. The second category includes applying models of ML algorithms, statistical learning algorithms, and deep learning algorithms. These software models try to learn through the distribution of the data and patterns which are directly extracted from the input data.

Many authors have been encouraged to create and deploy alternative prediction models as a result of recent developments in deep learning methods and the large quantity of the data accessible, as mentioned in more depth in the related work section. Despite this, the findings in the literature remain inconclusive [3]. As a result, air pollution forecasting remains a vital and active study field. The use of photographs for air pollution prediction is, in particular, an unknown topic that requires greater study.

Because of their influence on citizens' daily lives, high-quality projections on this subject are crucial. Knowing ahead of time of high-pollution occurrences can lead to regulations like traffic restriction, the closing of public places like schools, and suggestions to reduce exposure for sensitive populations. As a result, high-quality predictions need adequate detection approaches that are suited to the task at hand and can solve intrinsic problems such as handling class imbalance and utilizing multimodal data.

Assessment of air quality is the most significant part of the air pollution monitoring and control. The suitability of air supply for a specific application is determined by its qualities. Criteria air pollutants are a group of contaminants that are extensively spread across the United States. These pollutants might be the reason to harm the people living area, environment, people's health, and their property. The following pollutants have been designated as criteria pollutants:

- Ozone (O_3);
- Sulfur dioxide (SO_2);
- Nitrogen dioxide (NO_2);
- Carbon monoxide (CO);
- Particulate matter (PM);
- Lead (Pb).

The EPA, local, tribal air contamination or pollution control agencies, state, and use the air quality system (AQS) to collect data on ambient air pollution from thousands of sensors. AQS also includes descriptive information about each monitoring station (such as operator and its geographical location), weather data, and data quality/assurance control information data from the AQS is used to aid in Attainment or Non-Attainment designations, monitor air quality, do modeling for permit review analysis, analyze State Implementation Plans for the area which is in Non-Attainment, and execute other air quality management duties [4]. The Clean Air Act requires that AQS data be utilized to generate reports for Congress.

The standards differ because different contaminants have different impacts. There are guidelines for short-term as well as long-term average durations for some contaminants. Short-term standards were developed to protect against acute or short-term health impacts, whereas long-term standards were developed to protect against long-term health problems (Table 3.1).

Carbon dioxide (CO_2), nitrogen oxides (NOx), hydrocarbons (HCs) or volatile organic compounds (VOCs), PM, and carbon monoxide (CO) emissions from vehicles have become a major cause of air pollution in many areas. Increased traffic congestion, especially along major roads, has the reason to drastically degrade air quality and increase contaminant emissions. These emissions increase the risk of illness and death for drivers, commuters, and people who live adjacent to highways, according to epidemiological research, estimations of proposed vehicle emission standards (VES), and environmental impact assessments (EIA) for WHO's specific road projects.

This study explores the quantity of health risks and air pollution to near and on-road inhabitants that might happen as a result of repeated traffic congestion, such as weekdays (Monday to Friday) in rush hour traffic. Recurring congestion can lead to more chronic and frequent exposures, as well as an increase in long-term health risks [5]. Despite the fact that such occurrences may have a considerable impact on some acute health outcomes, such as asthma exacerbation, "the occurrence of overcrowding" such as that activated by a disabled vehicle (due to accident), is not addressed. Predictive risk assessment methodologies are used in this study, including emissions, traffic simulation models, pollutant dispersion, and its risk, as well as an incremental study or analysis that analyzes congestion-related and congestion-free outcomes. To investigate the effects of important factors on the predicted additional risk, a brief sensitivity analysis is performed.

TABLE 3.1 Potential Health Consequences of the Air Quality Index (AQI) [4]

AQI Value	Range	Color	Recommendatory
0–50	Good	Green	Simply least bother or nothing will be hazardous.
51–100	Moderate	Yellow	Individuals who are unusually sensitive to the outdoors should restrict their time outside.
101–150	Unhealthier for group of sensitive people	Orange	Excessive outdoor activity should be avoided by active adults, children, and people suffering from respiratory illnesses like asthma.
151–200	Harmful to health	Red	The prolonged outdoor activity must be avoided by active adults, children, and people suffering from respiratory illnesses like asthma; everyone else must limit their time spent outside for long periods of time.
201–300	Dangerous and unhealthy condition	Purple	The outdoor activity must be avoided by active adults, children, and people suffering from respiratory illnesses like asthma; everyone else has to limit their time outside for lengthy periods of time.
301–500	Very hazardous to human health	Maroon	Outside physical activity must be avoided by everyone.

Source: Adapted from Ref. [4]

The majority of air pollution approaches rely on unprocessed or processed data, such as pollutant sensory measurements from the past. The innovation of our study lies in the use of multi-modal data consisting of sensor measurements and camera pictures, distinguishing our study from earlier approaches in the literature. Combining all these features helps to address issues associated with the prediction job, such as the availability of class unbalanced data, incomplete, and possibly noisy information derived from a single data source, and vulnerability to adversarial assaults. Consequently, the resultant predictive model looks to be more robust and accurate than previously published solutions.

3.1.1 BACKGROUND AND RESEARCH MOTIVATION

Air pollution commonly occurs when dangerous and unwarranted/ unauthorized quantities of specific chemicals such as gases, particles, and biological molecules are introduced into the atmosphere. Excessive emissions have apparent repercussions, such as disease and mortality among populations and other living species, as well as crop damage. Liquid droplets, Hard particles, and smokes are all types of air pollution, and they are classified as follows. The combustion of fossil fuels from vehicles, as well as industrial, domestic, and agricultural sources, are all common contributors to air pollution. The critical concern air pollutants are nitrogen oxide, SO_2, and CO. They are all that are described as 'primary' pollutants since they are emitted directly into the atmosphere. Secondary pollutants are created in the atmosphere, primarily as a result of UV exposure to primary pollutants. Ozone and secondary organic aerosol (SOA) are common secondary pollutants. Ozone (O_3) is a secondary pollutant that forms in the environment whenever NOx and hydrocarbon are exposed to UV light [6]. The generation methods of SOA from fossil-fuel burning remain unknown, although it is a major component of Beijing winter haze pollution.

Air pollution causes a wide range of negative health impacts. For six of the most prevalent air contaminants, the EPA has set National Ambient Air Quality Standards (NAAQS). PM, lead, CO, ground-level ozone, NO_2, haze, and SO_2 are the pollutants in question. The occurrence of these pollutants in the circulatory air is owing to a variety of different and widely disseminated sources of emissions. The NAAQS have been established to standardize the safety of public health. In addition, EPA has also established additional NAAQS to standardize the safety of public health which may impact the eye impairment for humans and safety of animal rights. The AQI is used as an indicator to identify the air contamination location-wise and region-wise. CPCB recently introduced the breakpoint concentration method of AQI assessment, which is for individual pollutant AQI estimation followed by the maximum of these as a synergistic AQI level that can be used for decision making. Since the last quarter, the USEPA and CHINA have been using the notion of breakpoint concentration levels in their research and development [7]. The additional pollutants that are used to calculate the "good" indicator AQI of daily air quality that might harm people, such as air toxics have been already discussed in Table 3.1.

If we can standardize AQI to forecast the concentration and contaminants of air pollutants hourly, the standard for AQI will become more flexible and beneficial to the safety of public health. Consequently, these forecasting systems are of paramount importance to ensure that proper warning has been issued to the public. These forecasting systems can be useful for governance to alert their community about air pollution exceeding the threshold values. They could also be considered for inclusion in the community governance with their existing emission control programs [7].

3.1.2 REVIEW METHODS AND QUESTIONS

According to Kitchenham's work [8] and the PRISMA guidelines [9], we have conducted this review. As a result, the technique for conducting this review is described in the subsections below. Consequently, we drew inspiration for our review's design by referring the survey of dengue disease surveillance [10]. The following are the main research issues addressed in this review:

- **Q1:** How much has data mining technology and techniques been applied in the epidemiology of air pollution?

  To answer this question, we conducted the relevant literature survey with a focus on the epidemiology of air pollution and research works related to data mining tools, technologies, and techniques. We have not set a time-bound limit for this review, but recent work has been given more importance. We have also considered the growing trend of data mining tools, technologies, and techniques during this review.
- **Q2:** Are there any hotspots for research in the studied field?

  Our review focused on covering the review survey covering all aspects, regions, countries, and research institutions. We have also focused more on those countries who have more advances pertaining to our review.
- **Q3:** Identify the focus areas of air contamination in which Data mining has been used previously.

To resolve the issue of finding focus areas based on the question Q3, we have filtered through various categories of epidemiological work related to air pollution and its application areas. In the end, data mining tools, technologies, and procedures showed three types of epidemiological investigations of air pollution in the survey.

- **Q4:** Which data mining techniques were used?
  To resolve the issue of finding the data mining techniques and algorithms involved, we conducted a vast review and found various data mining techniques pertaining to air pollution prediction and found four major categories which we will discuss subsequently.
- **Q5:** What are the current work's limitations?
  To overcome the limitation and focus on resolving the issues, we classified all the issue techniques pertaining to air pollution prediction under a few categories and strategically tried to address all the major issues that occurred in the existing systems.
- **Q6:** What potentially fruitful directions haven't been pursued yet?
  Finally, in Q6, we discussed possible future actions.

This chapter has been organized as follows: Section 3.1 provides an overview of Air pollution and health risks due to vehicle traffic. Section 3.2 discusses relevant work that has been done to address the challenge of air pollution forecast and prediction. The architecture of the proposed models and the data preprocessing techniques are explained in Section 3.3. Section 3.4 deals with the ensemble learning (EL) based prediction. In Section 3.5, we provide the evaluation results of our models and discuss them. Finally, Section 3.6 concludes the chapter.

3.2 LITERATURE SURVEY

Yi et al. [11] proposed, a large amount of research was devoted to the prediction of air pollution, taking both standard and deep learning techniques into consideration. Even though many alternative models have been developed, they all rely on pollution data obtained from sensors at certain times and locations to some extent. The gaseous species and particulate matter with the values PM-2.5 and PM-10 are regarded pollutants are typically the following, i.e., CO, SO_2, O_3, and NO_2. Integrated into the system are meteorological data such as temperature, rainfall, humidity, and wind speed wherein few systems even incorporate weather prediction indicators information. Fan et al. [12], discussed the modeling and preprocessing phases are often carried out in distinct steps. An important stage in the data preparation pipeline is feature extraction. The regularly utilized strategies incorporate discriminant examination, factor investigation, group examination and head part investigation (PCA). Matching along with aggregation by location and time, together with extrapolation and interpolation techniques, are used to achieve data fusion.

Qi et al. [13] explained the stage of feature analysis or interpolation is performed separately, certain methods include this stage directly into the model design. Several 100 techniques have been published in the literature, and they may be divided into two main groups. The first group of techniques is concerned with forecasting the amount of contaminants present (such as NO_2, PM-2.5, PM-10, etc.). Kalajdjieski et al. [14] proposed the toxin and pollutant method arrangement into two fundamental gatherings. In the principal class of strategies, contaminations are anticipated by determining how many toxins and pollutants, for example, NO_2, PM-2.5, PM-10, and so forth. Because of such regulation and air quality proposals, European Union and other public ecological associations have fostered their own principles and rules for these poisons. Air quality is estimated by the AQI, which reflects how perfect or unclean the air is, just as any related well-being perils, particularly for in danger populaces. Health effects that may arise after a few days of exposure to polluted air are the subject of this study. In order to arrive at this number, the highest AQI values for each of the previously specified criteria pollutants are used.

Li et al. [15] proposed the AQI or air pollution index techniques, which are included in the second group of methods for forecasting the pollution level. The labeling of pollutant data is the most significant distinction between the two kinds of techniques. When it comes to supervised classification algorithms, the first category of methods does not require labeling because the goal is to predict actual pollutant levels. However, the second category of methods does require labeling fine-grained data because it will be used for supervised algorithms.

Corizzo et al. [16] proposed autoencoder models succession to-arrangement models neural organizations that incorporate direct indicators as troupes, Bayesian organizations, and multi-name classifiers are a portion of different strategies that have been utilized as of recent years. The AQI will be more responsive and better for the health of the community, if a prediction system based on the percentages of certain pollutants and susceptible of predicting air quality hourly-based is developed; because AQM systems are essential for communities. By letting environmental regulators to select between "on-demand" reductions in emissions, operational planning, and maybe even emergency action, they can play a vital role in health alerts when pollution levels near stipulated thresholds.

Liu et al. [17] discussed the attention-based model used in is another fascinating approach. As a result of this method's attention mechanism, just the wind measurements are used to generate an encoded result and then used that in the dominant framework. Using attention-based methods, we apply them to all of the available meteorological and pollution information. Wen et al. [18] discussed the approaches employ a variety of evaluation methods. They all use standard regression evaluation processes such as the relative prediction error (RPE), mean prediction error (MPE), mean absolute error (MAE), mean absolute percentage error (MAPE), and root mean square error (RMSE).

ARIMA is an abbreviation that represents the autoregressive incorporated moving normal model. For time series anticipating, it is one of the most significant and generally utilized models. Due to its factual properties, flexibility to clarify a wide scope of cycles, and extensibility, this model, at first proposed by Box & Pierce [19], had gained critical help in established researchers. Statistical techniques like ARIMA have become extremely prevalent for forecasting air pollution levels including air quality in metropolitan areas over time since the advent of public concern about urban area air quality and living quality. The quality of the data and its representativity are critical factors in ensuring the effective construction of forecasting models. The data preparation stage frequently has an effect on a ML algorithm's generalization capacity. Missing data are often imputations, outlier observations are removed, data transformations (such as normalization and standardization), and feature extraction are all part of the preprocessing of data. To produce an even information dispersion and to limit information inconstancy, the third stage is often as possible utilized related to the past two stages, which are helpful for getting more definite and complete informational indexes. Finally, a new dataset, frequently smaller but more informative, is generated in the fourth phase. The last advance step incorporates the extraction of elements and the choice of the main ones.

According to Siew [20], the capability of the proposed model is to estimate the outcomes for such AQI was investigated, with the outcomes uncovering that it was equipped for creating figures that fell beneath the 95% certainty level. Siew [21] proposed the level of self-assurance ARIMA's success in recent years has been particularly impressive. An exponential smoothing model was used to anticipate daily AQI levels based on findings. While classical statistical models have been used in time-series forecasting, ML models have gained interest as a solution that may be used to make precise forecasts in many scientific areas and spheres of endeavors. To predict air quality, in especially, ML algorithms have been widely used.

D'az Robles et al. [22] conducted an investigation on the deployment of a hybrid model incorporating ARIMA and ANN and to forecast the air quality in Chile, specifically P-10 data, using ANNs and ARIMA. ANNs were employed to reflect the nonlinearity observed in the ARIMA regression, allowing for a more accurate representation of the dataset's linearity. The researchers reasoned that the last model has incredible speculation capacity and outflanks both by using ANNs and ARIMA when utilized autonomously. Using datasets from other pollutant concentrations, Castelli et al. [23] designed an evolving framework system for estimating ozone concentrations using GPS up to an hour in advance. The strategy produced accurate findings that outperformed those obtained by state-of-the-art ML approaches. This approach has been employed by a number of researchers to anticipate air quality and pollution levels.

SVR has been examined with artificial neural networks (ANNs) in a study conducted by Muller et al. [24]. In general, the authors concluded that SVR performance was superior. A time-series forecasting method combining ANN as well as support vector machines (SVMs) to divide the input space into manageable chunks. This hybrid technique delivers great prediction performance while also allowing for efficient learning, as seen by the findings. Anikender & Pramila [25] adopted Multiple Linear Regression Techniques and Principal Component Regression and to anticipate the routine AQI value for Delhi, India, based on prior data of AQI and meteorological indicators. The Research work was funded by the Indian government. With the use of historical data from the years 2000–2005 and several calculations, they anticipate the daily AQI for the year 2006. Multiple linear regression was used to compare the 2006 AQI predicted value with the observed value of 2006 AQI for each of the four seasons (summer, post-monsoon, and winter). Principal component analysis (PCA) is a technique used to determine whether the independent variables are related or not. The principle components were used to minimize the number of data set as well as to eliminate covariance among the predictor variables while doing multiple linear regression. In the winter, Principal Component Regression is used to predict the AQI. AQI forecasts were based only on meteorological data, and no attention was given to fine particulate matter that might have a negative impact on human health.

Huixiang et al. [26] identified two separate towns, one in Beijing and one in Italy. In accordance with two publicly accessible datasets, have projected the AQI for Beijing and the quantity of NO_2 in Italy. As of August 2018, the Beijing Public Works Environmental Center has released its first dataset, which includes features including continuously aggregated AQI, NO_2, PM-2.5, PM-10, O_3 and SO_2 concentrations in Beijing. Between December to August 2018, this dataset has 1,738 occurrences. The second Dataset, which contains 9,358 occurrences, was gathered from an Italian city between March 2004 and February 2005 for the purpose of this study. The properties listed below are contained inside this dataset. CO, non-methane HCs, benzene, NO_2, and NO_2 concentrations on an hourly basis. However, they concentrated their efforts mostly on NO_2 prediction since it is the most important variables for air quality assessment. The approaches Random Forest Regression and Support Vector Regression methods were utilized to predict the AQI and NO_2 concentrations. SVR outperforms RFR in terms of accuracy in forecasting AQI, whereas RFR outperforms SVR in terms of accuracy in predicting NO_2 concentration.

The information on air pollution during the time period of 2009 to 2017 was examined by Nidhi et al. [27], who also provided a critical assessment of Delhi, India's air pollution trend through 2016 to 2017. Their predictions included SO_2, ozone (O_3) benzene, NO_2, PM-2.5, CO, and a range of other pollutants. Earlier executing data, utilizing Time-series relapse anticipating, they anticipated future levels of the pollutants indicated. The monitoring stations in the area AnandVihar and Shadipur in Delhi are under investigation as a result of the findings of this research. PM-10 concentrations have increased significantly, while NO_2 and PM-2.5 values have also risen, suggesting a rising degree of pollution in Delhi. A drop in CO, a rise in NO_2, an increase in ozone, a decrease in benzene, and a rise in SO_2 are all predicted for this year's air quality measurements.

Mohamed & Rakesh [28] evaluated the ratio of multiple air pollutants (SO, NO_2, CO, PM-10, and NO) by the span of day and week and dissect the impact of climatic factors like wind velocity, temperature, and dampness on the recorded air contaminations utilizing the WEKA Tool. The contamination control leading group of Karnataka gathered the information. The investigation discovered that when the ZeroR calculation is utilized related to the WEKA tool, the focus levels of air toxins increment during normal working days, especially at the hour of pinnacle hours of the day, and drop at the hour of ends of the week and excursions. The study illustrates the association or dependence between environmental variables such as humidity, wind velocity and temperature, and the air pollutants such as SO_2, NO, PM-10, CO, and NO_2 using simple K-means Clustering methods. In time series analysis, delays are used to quantify series autocorrelation which important phenomena that occurs in data sets. For forecasting, the selection of appropriate time delays and the removal of redundant information are crucial. A greater understanding of the underlying process may be gained by using this method in association with other forecasting techniques. The examination study was done to decide how much slack factors for every toxin and molecule through utilizing autocorrelation capacity and partial autocorrelation. First and second lag Partial autocorrelation graphs indicate gains and declines, respectively. With an 80% confidence level, only the first and second legs were judged important for all pollution series, leading to the creation of 10 additional variables. Finally, the final highlighted variables relate to the rolling mean of each pollutant series across a 24-hour time window.

Kazem et al. [29] employed the WHO-recommended AirQ software, which offers quantifiable amount of data such as SO_2, O_3, PM-10 and NO_2 on the health of residents in the city of Tehran, Iran. All-cause mortality, cardiovascular illnesses, and respiratory disorders were all considered in terms of health implications. Pollutant PM-10 previously ha d greatest impact on the city's populace is 8,700,000 and added to an all-out mortality of 2,194 fatalities out from 47,284 numbers in a year, as per the conclusion of the exploration. The total extra mortality cases attributed to Ozone, NO_2, and SO_2 are 819, 1,050, and 1,458, respectively. The findings indicate that the city has a serious air pollution problem and that the city's health burden from air pollution must be reduced.

Gunasekaran et al. [30] conducted a study to analyze the air quality in Salem city of Tamil Nadu state, India, during the time of April 2010 to March 2011. The yearly normal degrees of toxins like sulfur dioxide, nitrogen-oxide, and PM in this area are within the acceptable range of guideline, per the results of this study. However, the yearly average PM-10 value is somewhat higher than the national guideline. Additionally, the monthly average PM-10 concentrations exceeded the standard threshold with the exception of July to October.

3.3 SYSTEM ARCHITECTURE

The proposed system's framework is defined via architectural layout, which exhibits the practice of determining the required software, hardware, and associated interfacing components. It establishes the structure and behavior of the system and several other aspects from multiple perspectives. The system that has been designed is small and light enough to fit in any vehicle. As a result, dangerous gaseous concentrations from moving vehicles are mitigated through further prediction processes. It ensures a significant shift in air quality management and control strategies. This entire physical system is divided into two segments modules. The elaborated description of accounted modules are given in subsections.

3.3.1 MODULE 1: DATA PERCEPTION

In this study, SO_2, O_3, NO_2, CO, and Particulate Matter (PM10) are five specific air pollutants from vehicle emission. Pollutant emissions from vehicles are detected with the use of sensors. The sensors employed for experimental trials include MQ7, MQ131, Mic 2714, SO_2, and PMS5003. It's well known that MQ7 has a high sensitivity to CO, Mics 2714 detects NO_2, MQ131 senses O_3, and SO_2 sensor is utilized to trace SO_2 emissions, and finally, PMS5003 is utilized to monitor PM10.

The Raspberry PI 4 modal B board collects the sensing data and process it. The Raspberry PI has been configured to gather sensed data from several sensors and correlate it to predetermined standards to evaluate the accuracy. Five LEDs are incorporated to show the present state of various pollutants. The LED indicators turn red if the detected values are higher than the preset ones; otherwise, it shows green indications to exhibit that the readings are within the acceptable range. If the emission levels continue to be higher than the threshold,

Raspberry PI sends the information to the connected GSM module for further processing.

The GSM module provides detailed information and notifications to the users and traffic administrators as messages. An illustration of the architecture model is depicted in Figure 3.1. The Raspberry PI board is wired up with the sensors. Sensor data is continuously sent into the Raspberry PI device. A circuit connects the microcomputer to those LED indicators. An interface connects the GSM module to the board. The GSM component includes a cellular sim card.

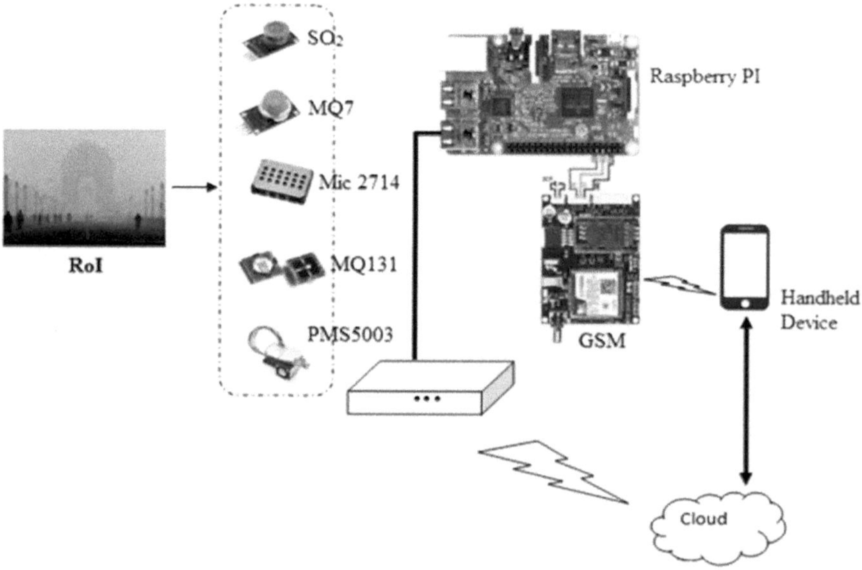

FIGURE 3.1 System architecture.

Each value obtained during a specific period from the Raspberry PI board is saved as a portion of the dataset. This assists us in keeping tabs on how the atmosphere is changing over time. The obtained information is utilized to build and train the models that perform predictions and analyzes on the complete construction of the dataset with sufficient values. The dataset is made up of real-time data gathered from a wide range of sensors in the field. Each model requires the extracted feature set as an input. The model uses a machine-learning algorithm as base predictor and an EL predictor model to refine the prediction accuracy using real-time and past datasets.

3.3.2 MODULE 2: DATA PREPROCESSING

We did research on New Delhi pollutant sites that are among the most polluted cities in the entire globe. Central Control Room for Air Quality Management (CCRAQM) provides the actual data. Data preparation is the first step in analyzing these readings. The learning phase is preceded by a number of preprocessing processes. If a variable is found to be incorrect, it'll not impact the entire data set. Instead, the parameter will be labeled empty or, if applicable, substituted with results calculated from the CCRAQM, avoiding the removal of the entire row. Imputation is used to retrieve the appropriate values when there are missing ones. The missing data regarding pollutants are imputed based on the measurements from the actual observation sites. For the remaining incorrect or incomplete information, the k-NN technique was employed to impute them using the prior imputation procedure. All the accounted datasets have a missing value/ incomplete information of less than 1.3%.

As a result, the factor relevance won't be polluted by range or unit effects. The input and intended data are then adjusted to remove any bias. The range of values for all of the raw data is set between 0 and 1. Measurements dominated by sources with a larger scale tend to be given more importance. This is because the normalization increases the pace of model learning while simultaneously supporting the k-NN computation results due to the distance metric being used to determine imputation. Predictive approaches use hourly AQI data with the maximum indices of five pollutants: O_3, PM10, CO, SO_2, and NO_2 to select the optimal features to include in the training and testing process. The accuracy of the present air quality evaluation will be significantly reduced if any of these variables are missing. The AQI (ψ) is computed based on Eqn. (1).

$$\psi = \{\{\left(r_{SO_2}, r_{CO}, r_{NO_2}, r_{O_3}, r_{PM_{10}}\right) \neq \varphi\varphi, Otherwise \tag{1}$$

Eqn. (2) transforms pollutant intensity value 'r' at the varying time, 't' into pollutant index 'I_p.':

$$I_p = \frac{\left(I_i^H - I_i^L\right)}{\left(\delta_H - \delta_L\right)}\left[\delta - \delta_L\right] + I_L \tag{2}$$

where; i denotes $(O_3 \| PM10 \| CO \| SO_2 \| NO_2)$, indicates AQI; δ denote pollutant concentration; δ_L concentration breakpoint at $(\leq \delta)$; δ_H

concentration breakpoint at (≥δ); and I_i^H, I_i^L are index breakpoint to δ_H and δ_L, respectively.

3.4 ENSEMBLE LEARNING (EL)-BASED PREDICTION

Instances of variables utilized for the prediction process include air quality parameters, weather data, and TIME (date/week/hour of 'nth' day), all of which may be found in the study. Data on air quality have been gathered from October 1, 2020, to May 31st, 2021. Graph neural network (GNN) and Random Forest are considered as base learners.

3.4.1 *GRAPH NEURAL NETWORK (GNN)*

This platform class uses graphs as test data and is known as a graph neural network (GNN). A class label is discovered by doing calculations on graphs. A single subjective component drives each recurrent unit in a GNN. Nonlinear recurrent modules receive inputs from both historical data and the results of current iterations.

The illustration in Figure 3.2 depicts GCN's general idea. The green node begins by averaging the mean values of its neighbors and its node ID. This value is then used to train a neural network on new information. Significantly, GCN makes use of a completely connected layer. Vectors in the second dimension tend to produce the desired results.

3.4.2 *RANDOM FOREST*

Using the bagging principle, another popular ML approach, random forest, mixes many decision trees to create a forest and randomization into the conceptual model. Individual trees are divided using a random set of features, while training sample subsets are created for every tree structure using a random distribution of samples. Once a node in a decision tree considers the optimum split for the randomized count of features. The algorithm will opt for the most prevalent result if the feature is categorical, whereas for numerical, the mean of total outcomes will be used.

Both regression and classification problems can be handled using random forest techniques. First, each testing data sample is run through the forest's decision trees multiple times to enhance prediction. Trees are then used to make predictions by voting on a resultant. The prediction is then generated by a consensus majority between the trees, leading to an improved unique learner. As predicted, random forests are better than choice trees since the prediction means closer to the actual truth as a categorization (regression). Figure 3.3 depicts a random forest made up of N random trees.

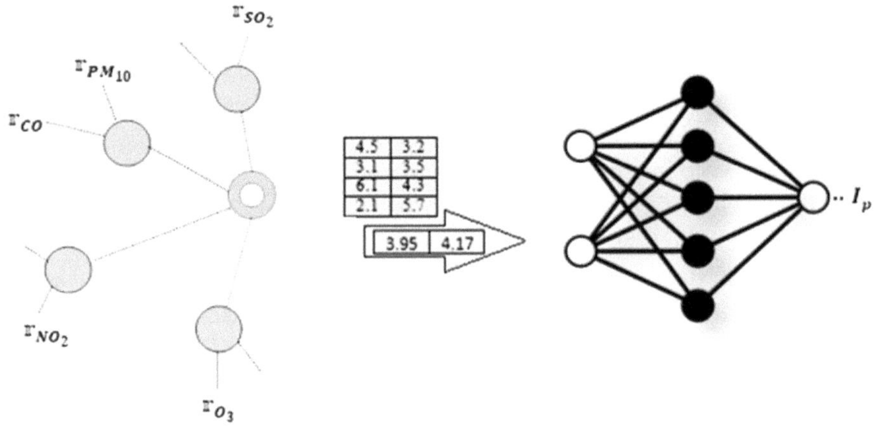

FIGURE 3.2 Overall concept of GCN.

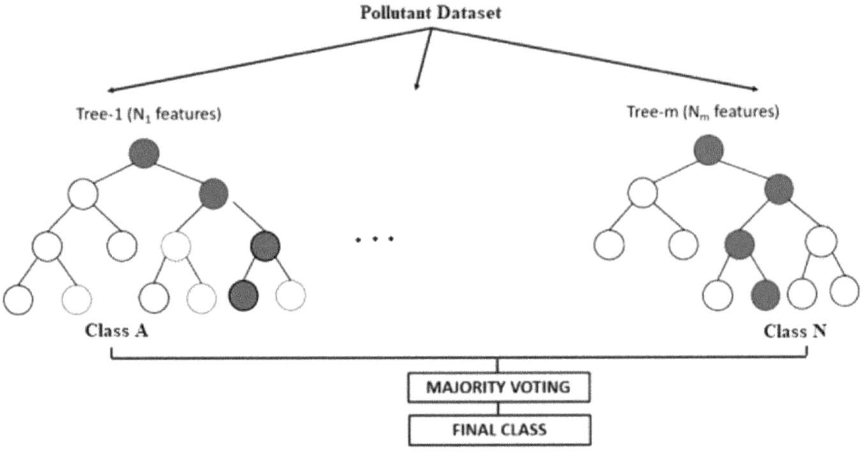

FIGURE 3.3 Overall concept of random forest.

3.4.3 ENSEMBLE LEARNING (EL)

This study uses EL, which combines a variety of learning and optimization models to get better outcomes than any one of them alone could. These types of training and learning mostly use dependency injection and parallel processing. It suggests a high degree of reliance among a small group of datasets that sequentially create results, using strategies such as boosting.

EL's core idea is to categorize or create results by using a variety of learning methodologies. Both voting and mean assessments may be used to combine the results of several classification algorithms. In linear or non-linear regression, data segmentation and mean assessment techniques are most often used, while voting methods are most commonly used in data labeling and classification. There are two sorts of voting: soft and hard. Classifiers vote independently, and a majority of those votes are authorized under the idea of "Hard voting," also known as "qualified majority."

Each classifier offers the probability estimates for one particular class label, one per data point; thus, soft voting may be regarded as a special kind of soft classifier. These probabilities are averaged out to determine which objective class label obtains the most votes. Many different ML based models may be combined into a single EL platform. It is possible that hybrid learning may improve the accuracy and resiliency of certain learning strategies, as well as overcome some of the limits of a particular approach. Basic learners, sometimes known as 'weak learners,' are among the incapacitated ones. 'Strong learners' are those who have achieved the baseline result and have built on it.

This section's goal is to look at and contrast the three various approaches that were previously described. An EL strategy based on soft voting was proposed in this chapter. Class labels may be deduced using the classifier's predicted probability p rather than the strict hard voting method. Baseline learners GCN and Random Forest are used in the initial phase of data processing. Once the accuracy of the baseline learners has been determined for each data point, the variances of those results are averaged. The EL forecast assessment is based on the soft voting results. Figure 3.4 depicts the full process.

3.5 RESULTS ANALYSIS AND DISCUSSIONS

There are two analysis parts. To begin, a brief overview of the dataset is delineated in Section 3.1. The data facts and information are primarily focused on geographical dispersion across New Delhi which is opted as RoI to analyze the study. Section 3.2 examines the performance of EL prediction models in depth once the parameters have been defined and imputation has been performed based on AQI classification described in Tables 3.2 and 3.3.

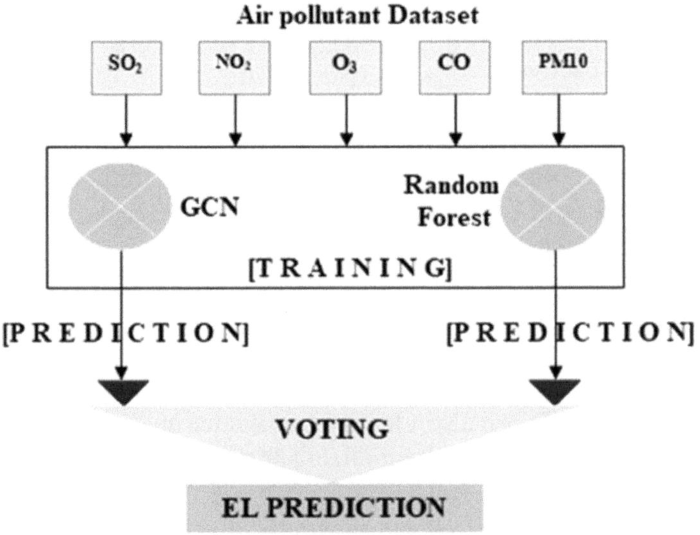

FIGURE 3.4 Overall process of EL strategy.

TABLE 3.2 AQI Category

AQI	Levels
0–100	Good
100–300	Moderate
>300	Unhealthy

TABLE 3.3 Analysis of Overall AQI Index Value

AQI Index	RoI					Overall Mean Index Value
	East Delhi	North Delhi	Central Delhi	South Delhi	West Delhi	
Unhealthy	43.5	45.625	42.5	47	48.375	45.4
Moderate	32.75	33	30.5	32.25	33.375	32.375
Good	23.75	21.375	27	20.75	18.25	22.225
	Average Index Value					

As a consequence of the observation period, moderate AQI levels are the most common from October to November. Conversely, unhealthy air quality is more common from December to April, showing that late winter and early spring are the most polluted seasons. From April to May, pollution levels generally reduce, with a modest spike in the early days of April. On average, more than 32.37% of instances fall into the moderate category, while 22.23% and 45.4% of cases go into the good and unhealthy categories, respectively.

Seasonal weather phenomena immobilize pollutants low to the land surface and keep them there longer in the winter, contributing to an increase in air pollution. As a result of pollution from industrial parks nearby as well as PM drifting in from neighboring states, the capital city has a reputation for having bad air quality.

Figure 3.6 shows significant instances of the unhealthy class (dark red bars) air quality for the month of April and May, with reduced pollution levels only in October and November. The worst air quality is concentrated from December to May (more than 60% unhealthy). The higher AQI levels in New Delhi are also clustered in winter and spring, but October and November also featured significant instances of the unhealthy class (respectively 35%, 38%, and 41%).

Root mean squared error and mean average error (RMSE and MAE) [19] commonly utilize key performance indicators that rely on the variation among a prediction's outcome and its actual value. In contrast, another measure, R^2 (R-squared), is vital to clarify the significance of the association between a predictive approach and its target output [20] (Figures 3.5 and 3.6).

(a) AQI category composition in North Delhi; (b) AQI category composition in South Delhi; (c) AQI category composition in East Delhi; (d) AQI category composition in West Delhi; (e) AQI category composition in Central Delhi

From the outcomes of Table 3.4, it's been noted that there are no significant differences in a random forest, GCN, or EL algorithm performance over the course of observational months. However, the R^2 performance of the random forest and ensemble algorithms is somewhat better. Over the period of all observation months, Random Forest performance lagged well behind that of the other two approaches, with the highest recorded values for assessment measures. The ensemble technique yielded an R^2 value of 0.764 as the best result. The MAE and RMSE averages of EL are 15.75%

and 26.38%, respectively. On the other hand, GCN outperforms random forest in terms of R^2 scores, notably in May, and accuracy metrics that are only marginally inferior to those achieved with RF and ensemble methods. As indicated, the findings are in line with those expected because as the prediction period becomes longer, the level of uncertainty rises.

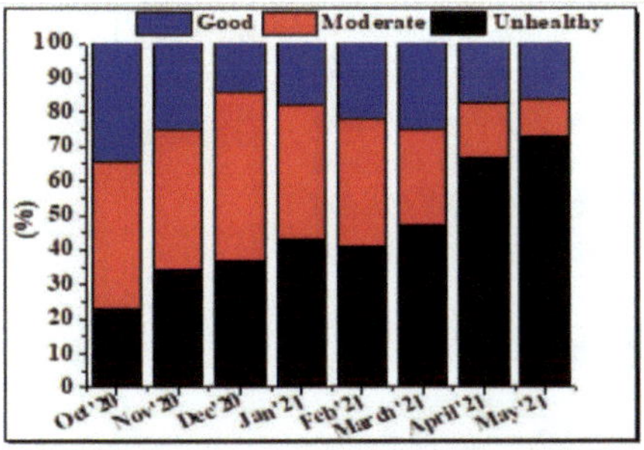

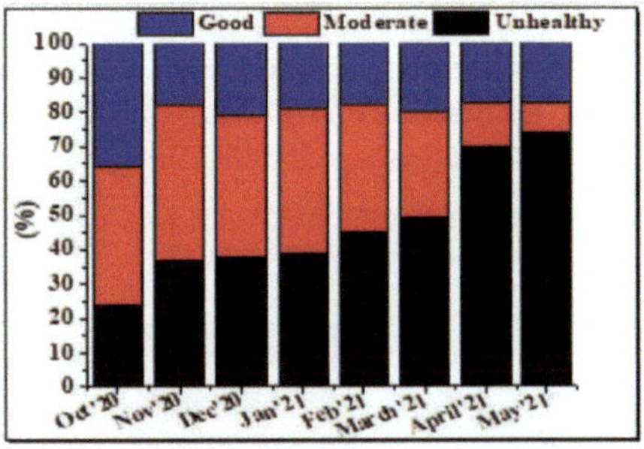

(a) AQI category composition in North Delhi

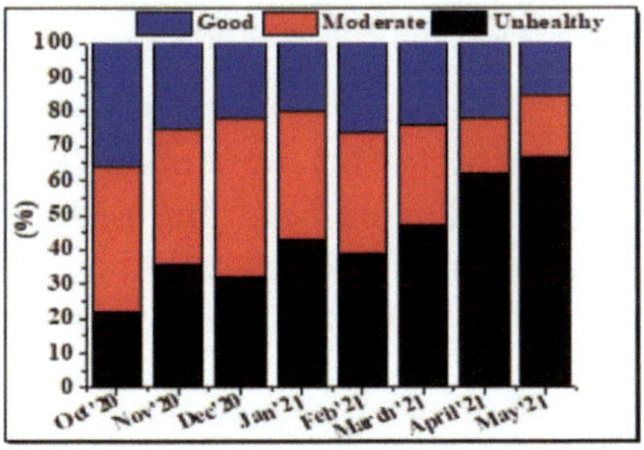

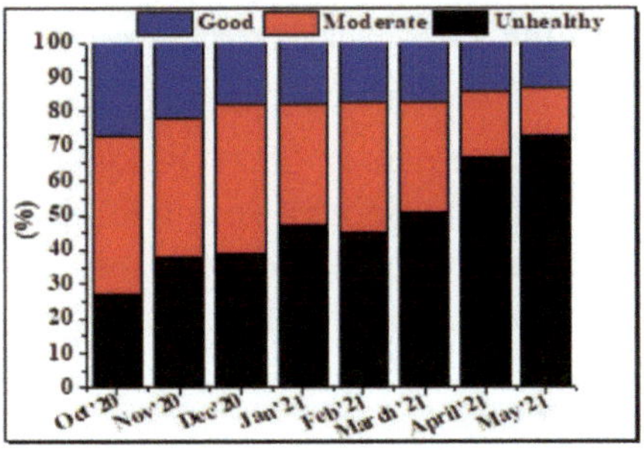

(b) AQI category composition in South Delhi

(c) AQI category composition in East Delhi(d) AQI category composition in West Delhi

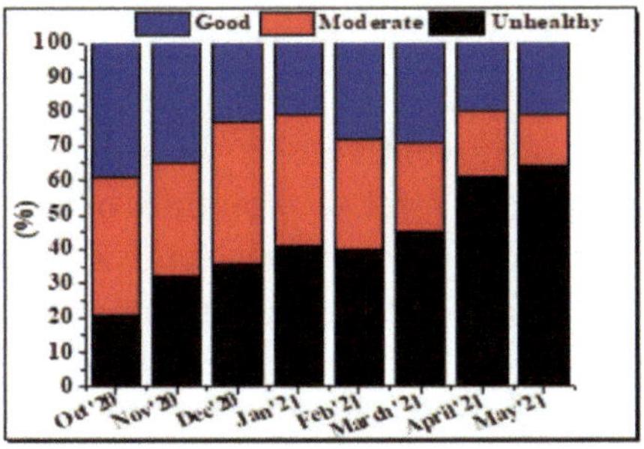

(e) AQI category composition in Central Delhi

FIGURE 3.5 AQI composition of RoI (month-wise).

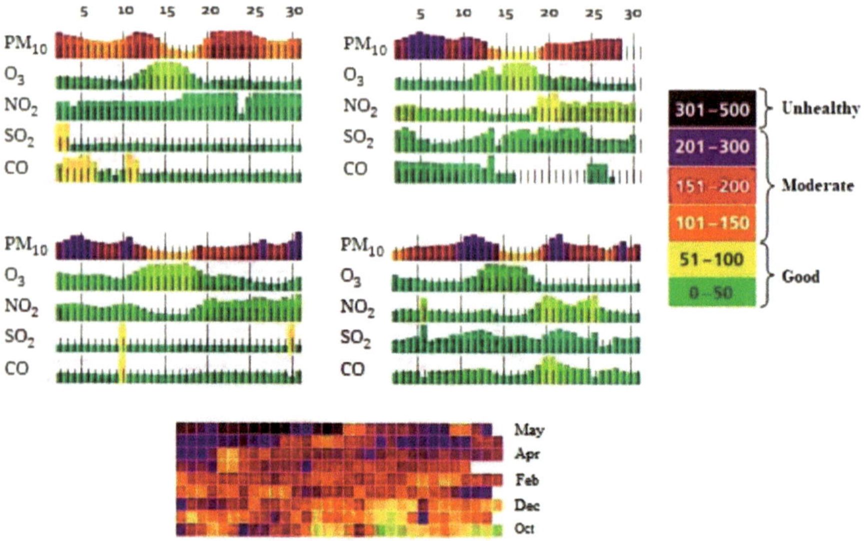

FIGURE 3.6 Pollutant recordings in RoI (month-wise).

EL surpasses other R-squared and RMSE in the investigation, but RF has the poorest result in all prediction situations. All the prediction resultants are shared with the traffic authoritarians to balance the traffic flow, which indirectly reduces the impact of pollutants on a particular area.

3.6 SUMMARY

There seems to be a surge in pollution throughout the last decades, resulting in several ecological problems. The use of artificial intelligence (AI) algorithms for AQI prediction has shown encouraging outcomes. Furthermore, the fundamental objective of the Smart Emission Surveillance System is to make it more inventive, user-friendly, effective, and timesaving than the present system. A total of 8 months of data were received from the CCRAQM's for training purposes, and a portion of the real-time dataset is constructed for testing and training purposes. The real-time datasets are perceived through the evolved smart air pollution monitoring system. Throughout the observation period, the quality of air in five parts of New Delhi (North, South, Central, West, and East) are considered for assessments due to their dire air quality conditions. The suggested method

TABLE 3.4 Outcomes of the Predictors

October' 20	Methods	Metrics				Methods	Metrics				February' 21
		RMSE	MAE	R^2	Accuracy		RMSE	MAE	R^2	Accuracy	
	GCN	24.13	15.42	0.341	89.23	GCN	27.51	19.67	0.462	85.46	
	RF	32.17	16.36	0.432	83.24	RF	35.78	21.43	0.327	89.18	
	EL	17.48	8.96	0.236	92.36	EL	13.78	14.26	0.271	94.37	
November' 20											**March 21**
	GCN	22.34	18.23	0.267	90.23	GCN	28.53	19.72	0.634	83.21	
	RF	35.11	18.92	0.478	80.34	RF	38.92	17.34	0.532	82.65	
	EL	19.23	10.12	0.194	94.36	EL	19.84	7.95	0.314	89.94	
December' 20											**April' 21**
	GCN	27.16	18.72	0.297	90.47	GCN	25.63	16.23	0.421	90.21	
	RF	37.32	15.33	0.523	87.18	RF	34.58	18.72	0.367	87.88	
	EL	18.03	9.74	0.281	91.56	EL	20.23	9.93	0.243	94.12	
January' 21											**May 21**
	GCN	28.19	18.62	0.435	88.42	GCN	21.22	18.34	0.729	86.17	
	RF	38.57	17.89	0.654	89.32	RF	28.93	19.26	0.782	83.56	
	EL	19.27	11.23	0.367	95.23	EL	19.2	15.63	0.479	91.23	

makes sensor outputs available via the internet for further inspection. With excellent outcomes for R^2, ensemble strategies provide the optimal target predictive accuracy based on the provided datasets. While the ensemble has better RMSE and accuracy, GCN has better MAE when used as a solo performer alongside the suggested one. This research also shows that the accuracy of the predictions differs by area across (India's) country's capital. The findings also show that when compared to other approaches, the EL techniques used in this work surpass them if examined. In this way, it may be seen as a distinct and better option for AQI prediction. In the future, we plan to extend the work on hybridizing the learning strategies with a large pool of datasets.

KEYWORDS

- **air quality monitoring**
- **ensemble learning**
- **Internet of things**
- **particle matter**
- **region of interest**
- **World Health Organization**

REFERENCES

1. Kalajdjieski, J., Zdravevski, E., Corizzo, R., Lameski, P., Kalajdziski, S., Pires, I. M., & Trajkovik, V., (2020). Air pollution prediction with multi-modal data and deep neural networks. *Remote Sensing, 12*(24), 4142.
2. Mishra, A., (2018). Air pollution monitoring system based on IoT: Forecasting and predictive modeling using machine learning. In: *International Conference on Applied Electromagnetics, Signal Processing and Communication (AESPC).*
3. Kang, G. K., Gao, J. Z., Chiao, S., Lu, S., & Xie, G., (2018). Air quality prediction: Big data and machine learning approaches. *International Journal of Environmental Science and Development, 9*(1), 8–16.
4. Patil, R. M., Dinde, H. T., & Powar, S. K., (2020). *A Literature Review on Prediction of Air Quality Index and Forecasting Ambient Air Pollutants Using Machine Learning Algorithms*, 5(8), 1148–1152.

5. Zhang, K., & Batterman, S., (2013). Air pollution and health risks due to vehicle traffic. *Science of the Total Environment, 450*, 307–316.
6. US Environmental Protection Agency (US EPA), (2015). *Criteria Air Pollutants.* America's Children and the Environment, US EPA, Washington, DC, USA.
7. Ramakrishnan, V., Chenniappan, P., Dhanaraj, R. K., Hsu, C. H., Xiao, Y., & Al-Turjman, F., (2021). Bootstrap aggregative mean shift clustering for big data anti-pattern detection analytics in 5G/6G communication networks. *Computers & Electrical Engineering, 95*, 107380.
8. Kitchenham, B., (2004). *Procedure for Undertaking Systematic Reviews.* Computer Science Department, Keele University (TRISE-0401) and National ICT Australia Ltd (0400011T. 1), Joint Technical Report.
9. Moher, D., Liberati, A., Tetzlaff, J., Altman, D. G., & Prisma Group, (2009). Preferred reporting items for systematic reviews and meta-analyses: The PRISMA statement. *PLoS Medicine, 6*(7), e1000097.
10. Runge-Ranzinger, S., Horstick, O., Marx, M., & Kroeger, A., (2008). What does dengue disease surveillance contribute to predicting and detecting outbreaks and describing trends? *Tropical Medicine & International Health, 13*(8), 1022–1041.
11. Yi, X., Zhang, J., Wang, Z., Li, T., & Zheng, Y., (2018). Deep distributed fusion network for air quality prediction. In: *Proceedings of the 24th ACM SIGKDD International Conference on Knowledge Discovery & Data Mining* (pp. 965–973).
12. Fan, J., Li, Q., Hou, J., Feng, X., Karimian, H., & Lin, S., (2017). A spatiotemporal prediction framework for air pollution based on deep RNN. *ISPRS Annals of the Photogrammetry, Remote Sensing and Spatial Information Sciences, 4*, 15.
13. Qi, Z., Wang, T., Song, G., Hu, W., Li, X., & Zhang, Z., (2018). Deep air learning: Interpolation, prediction, and feature analysis of fine-grained air quality. *IEEE Transactions on Knowledge and Data Engineering, 30*(12), 2285–2297.
14. Kalajdjieski, J., Korunoski, M., Stojkoska, B. R., & Trivodaliev, K., (2020). Smart city air pollution monitoring and prediction: A case study of Skopje. In: *International Conference on ICT Innovations* (pp. 15–27). Springer, Cham.
15. Li, X., Peng, L., Hu, Y., Shao, J., & Chi, T., (2016). Deep learning architecture for air quality predictions. *Environmental Science and Pollution Research, 23*(22), 22408–22417.
16. Corizzo, R., Ceci, M., Zdravevski, E., & Japkowicz, N., (2020). Scalable autoencoders for gravitational waves detection from time-series data. *Expert Systems with Applications, 151*, 113378.
17. Liu, D. R., Lee, S. J., Huang, Y., & Chiu, C. J., (2020). Air pollution forecasting based on attention-based LSTM neural network and ensemble learning. *Expert Systems, 37*(3), e12511.
18. Wen, C., Liu, S., Yao, X., Peng, L., Li, X., Hu, Y., & Chi, T., (2019). A novel spatiotemporal convolutional long short-term neural network for air pollution prediction. *Science of the Total Environment, 654*, 1091–1099.
19. Box, G. E., & Pierce, D. A., (1970). Distribution of residual autocorrelations in autoregressive-integrated moving average time series models. *Journal of the American Statistical Association, 65*(332), 1509–1526.

20. Siew, L. Y., Chin, L. Y., & Wee, P. M. J., (2008). ARIMA and integrated ARFIMA models for forecasting air pollution index in Shah Alam, Selangor. *Malaysian Journal of Analytical Sciences, 12*(1), 257–263.
21. Zhu, J., Zhang, R., Fu, B., & Jin, R., (2015). Comparison of ARIMA model and exponential smoothing model on 2014 air quality index in Yanqing County, Beijing, China. *Appl. Comput. Math, 4*, 456–461.
22. Díaz-Robles, L. A., Ortega, J. C., Fu, J. S., Reed, G. D., Chow, J. C., Watson, J. G., & Moncada-Herrera, J. A., (2008). A hybrid ARIMA and artificial neural networks model to forecast particulate matter in urban areas: The case of Temuco, Chile. *Atmospheric Environment, 42*(35), 8331–8340.
23. Castelli, M., Gonçalves, I., Trujillo, L., & Popovič, A., (2017). An evolutionary system for ozone concentration forecasting. *Information Systems Frontiers, 19*(5), 1123–1132.
24. Müller, K. R., Smola, A. J., Rätsch, G., Schölkopf, B., Kohlmorgen, J., & Vapnik, V., (1997). Predicting time series with support vector machines. In: *International Conference on Artificial Neural Networks* (pp. 999–1004). Springer, Berlin, Heidelberg.
25. Krishnasamy, L., Ramasamy, T., Dhanaraj, R., & Chinnasamy, P., (2021). A geodesic deployment and radial shaped clustering (RSC) algorithm with statistical aggregation in sensor networks. *Turkish Journal of Electrical Engineering & Computer Sciences, 29*(3).
26. Liu, H., Li, Q., Yu, D., & Gu, Y., (2019). Air quality index and air pollutant concentration prediction based on machine learning algorithms. *Applied Sciences, 9*(19), 4069.
27. Sharma, N., Taneja, S., Sagar, V., & Bhatt, A., (2018). Forecasting air pollution load in Delhi using data analysis tools. *Procedia Computer Science, 132*, 1077–1085.
28. Shakir, M., & Rakesh, N., (2018). Investigation on air pollutant data sets using data mining tool. In: *2018 2nd International Conference on I-SMAC (IoT in Social, Mobile, Analytics and Cloud)(I-SMAC) I-SMAC (IoT in Social, Mobile, Analytics and Cloud) (I-SMAC), 2018 2nd International Conference* (pp. 480–485). IEEE.
29. Tikhe, S. S., Khare, K. C., & Londhe, S. N., (2013). Forecasting criteria air pollutants using data-driven approaches; An Indian case study. *IOSR J. Environ. Sci. Toxicol. Food Technol, 3*, 1–8.
30. Gunasekaran, R., Kumaraswamy, K., Chandrasekaran, P. P., & Elanchezhian, R., (2012). Monitoring of ambient air quality in Salem city, Tamil Nadu. *Int. J. Curr. Res., 4*(3), 275–280.

CHAPTER 4

Various AI Models to Mitigate Transport Pollution

T. JANANI, P. SOBIYAA, K. SATHIS KUMAR, S. NAVEENA, and P. PONNILA

Assistant Professor, Bannari Amman Institute of Technology, Sathyamangalam, Tamil Nadu, India

ABSTRACT

In recent decades, vehicle usage increased enormously, and it made significant adverse effects on human health and the environment. Contaminants released by sources of transport cause major damage to the surroundings. Human action, especially the burning of fossil fuels that emit carbon dioxide (CO_2), methane, and other greenhouse gases (GHGs) into the atmosphere, affects the environment. Due to the transport pollution earth's temperature increases and that leads to global warming. Not all vehicles have similar effects. Transport pollution majorly occurs by burning the oil resources in the world and providing dangerous effects on the environment.

AI is making more changes in all kinds of sectors. Instead of mentioning it as changing, it is improving the current status of all technologies. AI shows a lot more improvement in the transport sector. The transport field has already adapted AI in several methods, and it provides autonomous functions for cars, trains, ships, and airplanes. Due to new methodologies of AI, even the traffic flow becomes smoother. AI-based autonomous transportation might minimize the human mistakes that cause so many traffic accidents.

Artificial Intelligence for Future Intelligent Transportation: Smarter and Greener Infrastructure Design. Rajesh Kumar Dhanaraj, Nilayam Kumar Kamila, Subhendu Kumar Pani, Balamurugan Balusamy, & Vani Rajasekar (Eds.)

People care about the water consumed by them but less care about the quality of the air they breathe. Day by day, the quality of air is being reduced due to the emission of pollutants. India's capital Delhi is popular for politics, protest, and highly for its pollution. The Central Pollution Control Board predicts severe air pollutant presence in Delhi and neighboring cities. The air pollution issue in Delhi has become a serious concern as the air quality index (AQI) continues to worsen. At noon, the town reported an AQI of 381. In the mid-week in Delhi AQI will be on average of 395. Innovative AI systems are launched to predict the quality of air. Systems such as the air quality micro-climate monitoring system (MCMS) are used to measure the quality of air and provide uncompromising data to the Government as a valuable resource to take appropriate action from the provided data.

Artificial Neural Network has been used to predict the contaminants which are present in the air using methods like multilayer perceptron, extreme learning machines, echo state neural network, regression model, and genetic algorithm (GA).

Various other artificial neural networks (ANNs) mathematical simulation methods such as regression, multiple regression, and time series analysis have been used to measure the air quality, reduce traffic congestion and produce efficient data. By minimizing traffic, transport pollution problems are addressed in the maximum concern. AI not only aims to make human lives easier, but it also makes transportation safe, clean, smart, and even better. AI is growing up to be one of society's most useful technologies. With it, we will begin to make major improvements to our soil, climate, and water and aircraft clean them up for a cleaner world.

4.1 POLLUTION

Pollution releases toxic, dangerous, unsafe substances into the environment and spoils its natural condition. Those substances are called pollutants. The pollutant may be in any form of gas, solid or liquid and some are natural pollutants like volcanic ash, hurricane dust, but the majority of the pollutants are created by human activities. Human beneficial product production is carried out in many industries and that causes heavy pollution in the environment.

The human activity involved in producing pollutions:

- Trash;
- Factory extracts;
- Vehicle emissions;
- Import-export activities through ships, airplanes, lorries, etc.

Humans pollute the environment in several ways. Vehicle usage has become huge because of the increasing population all around the world. Vehicles such as cars, motorbikes, buses, trucks need fossil fuel as intake to move from one place to another. Electricity is considered a safe and clean source of energy while generating electric power plants emit dangerous pollutants that severely affect human health. Most industries release waste into the air and water and those wastages cause damages to all living beings. Pesticides are used to fertilize the plants and to give more yield quickly. Pesticides have to be used at a very minimum rate but to prevent insects' high toxic pesticides are used. When those fossil fuels are burned, it emits terrible pollutants (Figure 4.1).

Pollutants
1. Particular matter
2. Volatile organic compounds
3. Carbon monoxide
4. Sulfur dioxide
5. Greenhouse gases
6. Nitrogen oxides

FIGURE 4.1 Pollutants.

Compared to the rural area, urban areas are more responsible for all kinds of pollution even pollution spreads to places where there are no lives. Pesticides and other chemical substances are being found in our soils present in Antarctic ice sheets. Plastics used by humans are not disposed of properly, resulting in the accumulation of plastics in the northern Pacific Ocean as a bunch. Not only it is contaminating the oceans with a long day-by-day air quality, but sea creatures are also consuming plastics and other waste substances as their food and losing their lives [3]. In factories during the manufacture of the products, toxic gases are released as the reaction of the process. Toxic gases mix into the air and spread all around the earth. When humans inhale the polluted air, the possibilities are very high to get a respiratory disease.

Every organism that lives in the world depends on the air, land, and water which is available as natural resources in the world. All kinds of organisms are affected when these resources are polluted.

4.2 TRANSPORT POLLUTION

One of the biggest challenges the universe is facing is pollution. There is a lot of pollution which the world is facing, among such one is transport pollution. In the modern era, automobiles have become of paramount importance to humankind. Everyday involves the use of vehicles by humans to reduce his traveling effort. For moving from one place to another transportation is involved which categories various modes like land, sea, or air. The most commonly used mode of transportation is by road which involves Two-wheelers, four-wheelers, and even Earthmovers.

Two-wheelers have become very important in the means of transportation where people use them for daily commuting to schools, colleges, offices, and business as well. Because of this frequent use on a large scale, two-wheeler manufacturing has increased a lot. The engines used in two-wheelers are 2 strokes and 4 strokes. Both stroke engines use fossil fuel which is extracted from earth mainly undergo various stages of refining which finally leads to the product like petrol and diesel which is the main source for automobile.

For transportation or moving goods from one place to another place on a large scale, humans are using heavy vehicles which mainly use fossil fuels as their source for running the engines. Day-to-day necessary items include food items, milk, and groceries mainly depend upon transportation as their factor to get moved from manufacturer to consumers. Because of modern civilization, humans started consuming foods from various countries which mainly needed goods to get transported in large numbers.

Trucks play a vital role in the movement of goods from one place to another place which uses internal combustion engines as their major source. These engines use petrol and diesel as their source and it emits carbon monoxide (CO) as residue after they get combusted. Not only does CO_2 properly exist. Day-by-day air quality sulfur dioxide (SO_2) and nitrogen dioxide (NO_2) which is very dangerous to humankind and nature as well. These gases cause harmful effects to lungs for human beings when it is inhaled, it blocks the transportation of oxygen to the brain [12].

Once these harmful gases are inhaled, it shortens the lifespan of humans by giving various side effects. Furthermore, these gases cause depletion in the ozone layer which in turn gives way for entering harmful ultraviolet radiation. This radiation causes various types of cancer to humankind. One such harmful disease is skin cancer. This radiation causes global warming on a large scale, and it will soon make the earth not fit for humans to live.

4.2.1 IMPACT OF TRANSPORT POLLUTION IN AIR

Pollutants produced by burning petroleum products which are produced from automobiles, airplanes play a vital role in air transport pollution. Incomplete combustion of petrol and diesel, when mixed with air in the atmosphere, leads to the formation of harmful pollutants like CO, carbon dioxide, and NO_2. The majority of this pollution is caused when the vehicle's engine is left idle in ON condition in traffic.

Nowadays, all people own two-wheelers and four-wheelers for their use and commercial use, which leads to increases in pollution when vehicles stop during traffic signals, leading to more emission of pollutants. CO produced from engines will cause direct poisoning when inhaled. This leads to liver damage and even to cancer when exposed for a longer time.

On the other hand, greenhouse gas (GHG) emissions from airplanes and Jet planes cause serious emissions of CO_2. The automobile and aerospace sector alone is not a major contributor to Air transport pollution. Major pollution is caused by emissions from Industries with a larger scale of pollutants, these pollutants will directly mix with air which reduces the Air quality majorly and leads to impure air [3]. The majority of industries including thermal power generation and steel manufacturing industries cause a major impact on Air transport pollution. These pollutants, when mixed with air, will cause acid rain and they will be mixed with normal rain and cause the worst effects on all living organisms.

4.2.1.1 ACID RAIN

Marble and other types of stone are also destroyed by acid rain. It has annihilated the words on gravestones and caused damage to several historic structures and monuments. Taj Mahal in Agra is well known for its white marble; it has become pale as a result of years of exposure to acid rain (Figure 4.2).

FIGURE 4.2 Acid rain.

Acid rain has been attempted to be avoided by producing energy that is required for human activity without the participation of fossil fuel. Some developed countries tried various methods to eradicate acid rain, but those methods would be a temporary solution. Renewable energy is the best alternative for air pollution because when used, it produces fewer pollutants when compared to non-renewable energy.

4.2.1.2 GREENHOUSE GASES (GHGS)

Greenhouse gases (GHGs) have a major contribution to the current level of air pollution in the world. GHGs like Carbon dioxide, Methane, present in the atmosphere, and also those gases have benefits too. GHGs are not completely a dangerous gas that causes air pollution, in case some amount of GHGs are necessary for the atmosphere. Plants breathe Carbon dioxide. Mainly to maintain the atmosphere in a proper warmth condition GHGs are essential [21].

Sunlight reaches the earth and so that all living organisms can survive. The heat of the sunlight remains in the atmosphere because Carbon dioxide absorbs the sunlight which is reflected from the earth's surface. Human lives are completely dependent on the warm sunlight we receive every day. This process of absorbing the warmth of the sunlight instead of leaving them into space is called the Greenhouse effect.

Forest trees are destroyed for many reasons but the primary reason is occupying the space for construction purposes. Vehicle usage among people increased very high in recent years, so burning fossil fuels became uncontrollable. Cut down trees and fossil fuels emission in the atmosphere are the major causes of the pollution in the atmosphere. GHGs automatically increase the temperature of the earth and become hotter than before and finally, it leads to an increase in the greenhouse effect. As a result, the greenhouse effect has risen, as have global average temperatures.

4.2.1.3 GLOBAL WARMING

Ice sheets and glaciers are melting as a result of global warming. Sea levels increase every year due to melting ice. Rising sea levels would gradually invade low-lying coastal areas. Climate change poses a danger to whole nations, such as the Maldives' islands. The phenomenon of ocean acidification is also intensified by global warming. Ocean acidification is responsible for many sea creatures' death. Ocean water absorbs some of the GHGs, due to that ocean water's temperature becomes high. Warmer, less salty waters may support fewer species.

The ocean water was contaminated heavily and the saltwater was converted into acidic water. This acidic water created a terrible condition for the survival of sea organisms, even the plants in coral. Scientists expect that as the world warms, extreme storms may become more common. Droughts will become more common in some areas, while flooding will become more common in others.

Pollution has already caused severe damage to the environment. The numbers of plants and animals are shrinking due to the effect of contaminated air, water, and land. Pollution spreads rapidly and is now found all over the world, including in the Arctic. Polar bears are starving and dying because their prey seals are no longer available due to contaminated ice sheets.

4.2.1.4 CHLOROFLUOROCARBON

Sun emits harmful radiation known as UV rays or Ultraviolet rays to the earth. UV rays from the sun will increase the risk of skin cancer and have a negative impact on the human immune system. Furthermore, these rays are harmful to almost all organisms, plants, and marine animals. Lives on earth are protected from the sun's harmful rays by an invisible pale blue protective shield called the ozone layer (Figure 4.3).

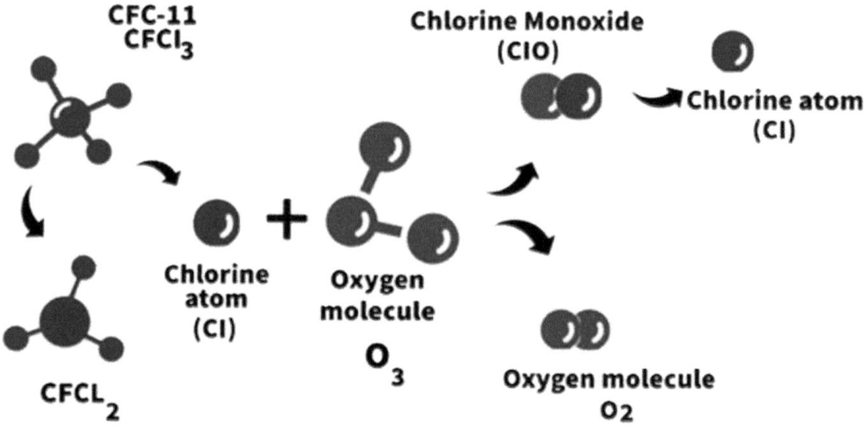

FIGURE 4.3 Breaking process of oxygen molecules.

Sun's UV rays are harmful, but those rays are consumed as energy by Oxygen molecules and transformed into a single atom of Oxygen. Oxygen atoms are bound together to form the Ozone layer. This surface, which is three to five millimeters thick and covers the earth, is three to five millimeters thick.

In the 1960s there was no Chlorofluorocarbon involved in the atmosphere. The ozone started to deplete after the 1970s due to the invention of harmful chemicals. When those chemicals start to emit and mix with other gases in the atmosphere then those emissions affect humans, animals, plants, and finally, it reaches the ozone layer and starts to deplete it. Chlorofluorocarbon is one of the Ozone depletion substances. The ozone layer, an area in Earth's upper atmosphere, is affected by CFCs. Once the ozone layer is affected, it causes major problems for the environment and humans. When the sun's UV rays contact the human body directly, they can cause major skin disease, damage the eye retina, and lead to a range of other health problems.

Climate change is also a cause of pollution in the environment. Also affects the ecosystem's physical and biological components. Furthermore, ozone depletion, GHG emissions, and global warming are mostly causes of environmental pollution. People and governments could reduce air pollution easily and efficiently.

4.3 IMPACT OF MARINE TRANSPORT POLLUTION

Billions of dollar goods are transported by ship every single day. About 90% of goods are transported through Marine freight globally. Global Marine transportation has positive impacts on economic development but causes negative impacts on the environment. The shipping sector is responsible for >3% of carbon dioxide emissions. The carbon dioxide emissions from shipping are expected to hike 50 to 250% by 2050 – estimation of International Maritime Organization (IMO). This will affect global climate change. Several hazardous gases such as nitrogen oxides (NOx) 15% per year, sulfur oxide (SO_x) 13% per year, hydrocarbons (HCs), Particulate matter, CO, and carbon dioxide CO_2) 2.2% per year. Gas emissions data can be tracked by the fuel oil consumption by marine engines. As gas emissions by engines are directly proportional to the total fuel consumption. Operational oil pollution is also one of the factors to be considered.

4.3.1 OIL SPILLS

Among various transportation pollutions caused due to several factors, the oil spill is a major factor that brings a huge impact to the environment. An oil spill in the marine ecosystem during transportation is also a form of pollution caused to the environment. Oil spills not only occur in marine life but also occur in the land in many forms. Example: crude oil and petroleum products from tankers, oil spills offshore platforms. Oil spills also occur during loading and unloading in port. All petroleum products are transported as cargo in ships. In case of any accidental discharge of oil due to any cause of the error, it will result in environmental disaster. Due to the physical and chemical properties of the fuel-efficient aircraft short-haul medium-haul oil, if spilled, it will undergo weathering, dissolution, oxidation, and volatilization, which creates a huge impact on the environment. The cost of cleaning up oil spills is lower than the cost of preventing small oil spills. As the cleaning cost is low, it does not have a huge financial impact.

However, it is a difficult task to clean them up, and it may take weeks, months, or even years. It stays in the same environment for years as sediment. Marine spice's organs and reproductive cycles are highly damaged due to the contaminants like polycyclic aromatic hydrocarbons (PAHs). Around 30 years ago, people witnessed a major disaster in Alaska. Lakhs of seabirds and thousands of sea otters and a lot more fishes lost their lives due to oil spills. Even though it does not impact humans directly, it indirectly affects those who consume seafood.

4.3.2 BALLAST WATER POLLUTION

Ships use a huge amount of ballast water to stabilize themselves when they travel without cargo. Cargo ships are stable when filled with cargo, empty cargo ships use ballast water when unloaded with cargo to balance the weight and to stabilize themselves. Marine transportation is a continuously growing sector and predicted to have rapid growth in the next few years. Around 3–5 tons of ballast water is discharged by ships annually. The discharged water is contaminated with oil and other impurities in the ballast tank. This is also called cruise pollution. This affects the marine species, vegetation in the sea, and sea animals as well.

4.3.3 AIR POLLUTION

Air pollution is caused by various sources, marine transport is also one of the factors that cause air pollution. Air pollution kills about 4.2 million people, says the World Health Organization (WHO). Most of the ships are powered by combustion engines which emit hazardous gases during their operation. It mixes into the air and causes various health issues to the living organisms. Similar to road vehicles, diesel, and other engines used in ships emit unwanted toxic air through the exhaust called vessels. The major ingredient of the emission is GHG (CO_2). Despite small passenger ships, cargo, bulk carriers, oil ships contribute the major emission of toxic gases to the environment [12].

4.3.4 NOISE POLLUTION

Noise pollution is also caused by ships. Ships use loud horns to indicate the presence of their ships to the other ships traveling and also to communicate with the port. Diesel engines and other oil engines also produce noise. Even electric generators used in ships also produce noise. As it is designed in such a manner that the sound will travel for a long distance. This affects the sea species which use sound waves to communicate with each other and the birds that fly to other countries overseas which use sound waves for their navigation also get affected.

4.3.5 SEWAGE POLLUTION

Large cruise ships can accommodate more than 7,000 passengers in which they travel for a long period. They also dump a million gallons of wastewater into the sea per week. Washing vessels that have dirt, showers, and laundries which contain toxic contents are some of the sources of dirt water. It is named gray water. Cruise ships also discharge harmful chemicals from batteries, chemicals from cleaning the daily utilities, chemicals from expired medicines. This creates a huge threat to the living organisms in the sea. Human wastes from toilets are also deposited into the sea by cruise ships. This form of waste disposal is classified as black water. This affects the marine ecosystem. Bile water which consists of highly concentrated oil that is leaked from the ship engines also pollutes the seawater.

4.3.6 SOLID WASTE POLLUTION

Solid wastes like plastic covers, steel cans, packing wastes, glasses are thrown off into the sea during travel. They float on the sea surface or deposit in the sea as marine debris. In case of any accidents during the cruise, the ship crashes in the mid-sea, and the materials in the ships are deposited as solid waste inside the sea. Ship breakage is also a form of marine pollution. Sea animals like whales and dolphins get affected by the solid waste deposited in the sea.

4.3.7 PHYSICAL DAMAGES TO MARINE

Cruise ships that travel at high speeds cause damage and injuries to the species in the sea. Huge whales strike and get struck in huge ships which causes marine damage. Tins and cans are thrown into the sea deposit above the coral reefs and destroyed. Anchoring of ships and ship accidents also disturbs the growth of the coral reefs.

4.4 ENVIRONMENTAL IMPACT OF AVIATION INDUSTRY

The aviation industry is expanding due to increasing passenger demand and also air freight. Big giants in the e-commerce industry offer same-day delivery which comes possible by air freight. Air mode of transportation is the fastest transportation system so far in human invention. Air transport can deliver the goods from one end of the country to the other end within a single day. The same will take a while if using other means of transport (road or ship).

Aviation is the most difficult sector in the transport industry to decarbonize. The aviation industry has taken many actions to make the aviation industry go green. Many kinds of research are going on to use biofuels as aircraft fuel. And also implemented in some of the airlines. Most Airlines are using Biofuel blends which are a mixture of bio and jet fuel and implemented in many commercial aircraft like Virgin Atlantic, United Airlines, and Qantas Airways. Anyways only some of the aircraft use biofuels. It takes time for implementation because of the scale of the industry and also huge investments, government support is needed to implement it in airlines. This would indirectly lead to an increase in ticket fares. There is a 32% increase in CO_2 emissions from commercial aircraft. As the aviation industry is growing faster than expected, the CO_2 emissions would triple in years. Around 5 billion passengers traveled by air in 2019.

The United Kingdom, Germany, The United States, Japan, and China are the top 5 countries that contribute most of the CO_2 emissions in the world. Big players in the aircraft industry have already started manufacturing fuel-efficient planes in which the emission of CO_2 will be reduced. The cost of producing, creating biofuel is high compared to normal jet fuel which creates a break in implementing biofuel. Not only biofuel is a solution there are also many interesting and efficient solutions too. Hydrogen-powered electric planes use hydrogen to power up the electric motor. Hydrogen-powered planes emit only vapor to the environment. And the other emerging solution is electric aircraft that use Lithium-Ion batteries to power up the electric motors in the plane. But it limits the endurance of the plane due to the high weight of the battery compared to fuels.

Future Airplane technologies like supersonic aircraft will burn fuel more than 5–7 times when compared with conventional aircraft, which are super polluting technologies. The aviation industry impacts climate change as well. Other than CO_2 airplanes also emit nitrous gases, water vapor, soot, sulfate, microparticles which lead to roof the factors like climate change, ozone formation, and decrease in methane, sun screening, creation of contrails, and creation of cirrostratus clouds of ice.

There are two major pollutions caused by the aviation industry which are discussed in subsections.

4.4.1 AIR POLLUTION

The air enters through the inlet of the aircraft engine and then enters into the combustion chamber combustion of fuel takes place. During the combustion process, it emits some toxic gases like CO_2, Nitrogen oxide, NO_2, and sulfur. They are emitted to the environment through the outlet of the aircraft engine which pollutes the air. During aircraft take-off, the engine works at higher efficiency which pollutes the air in the area around the airport regions. The major problem here is aircraft emit CO_2 and other toxic gases at higher altitudes that will have a greater and direct effect on climate change. Accelerated growth of the aviation industry resulting in the emission of 1.2 billion tons of CO_2 to the environment every year which increases thrice if the same situation continues and no action has been taken.

The major component of aircraft emission is CO_2 which is around 70% of the exhaust. Aircraft emits 3.16 kg of CO_2 for 1 kilogram of fuel consumption. It takes years to remove the emitted GHG from the environment naturally. The Stratospheric ozone layer is destroyed and global warming is caused by the NOx emitted by the aircraft. Aircraft emissions also result in chemical reactions in the atmosphere. Contrails are water vapors from aircraft engines that use fossil fuels and biofuels. At higher altitudes temperature is cold which leads to the freezing of water vapor which produces small ice crystals. The small ice crystals observe water vapor and they form as contrail-included cirrus clouds. It does not stay in the atmosphere for a long time, but contrails are also a factor that needs to be considered because contrails and cirrus clouds block the infra-red rays, which produce a warming effect [17], which is more dangerous than the emission of CO_2.

4.4.2 NOISE POLLUTION

Aircraft engines have larger blades inside them which rotate at higher RPM, which produces noise. Inside the aircraft's compression chamber, the air gets compressed and the air is released at high pressure and speed which produces a large amount of noise. It disturbs and causes some serious health issues like sleeping problems, psychological, and cardiac problems to the people who visit near airports. It's totally under the responsibility of the airport operator. Many measures have been introduced to reduce the noise during flight. Many restrictions on night flying and restrictions in flight paths have been implemented. Every airport should have the mechanisms to monitor sound and follow the restrictions. Noise limits have been given to some of the airports which are present inside the city and these airports will be under serious monitoring. Any operations beyond the limit are identified by the airport authority and the operator will be subjected to serious penalties. Due to increasing air traffic, it is tough to control and monitor the activities which lead to noise pollution.

4.5 ARTIFICIAL INTELLIGENCE (AI)

Technology is advancing faster and taking less time to be widely adopted than ever before. Artificial intelligence (AI) is one of the emerging technologies which have created huge impacts in various sectors. ML, computer vision, Algorithms, Big data, and AI are not just words but the future. People think that AI is replacing humans, but it enhances humans. There is a limit where humans cannot go or can be explained as humans cannot do. AI breaks through all the limits and it has the power to go beyond limits.

AI uses computer vision to sense the environment and ML to create patterns that could work out. AI combined with ML, computer vision can be possibly used for many applications in several sectors, which would result in making the "IMPOSSIBLE" possible. AI is also used to detect and reduce environmental pollution, which is also particularly used in the field of transportation. Over 30% of GHGs emissions are contributed to the environment by the transport sector. This will cause serious effects to humans and the environment, if the same situation continues and if a remedy is not found to reduce the percentage.

AI is growing in various fields even in transportation, where petrol and diesel cars are replaced by electric cars. Cars can now drive themselves in the real-time environment without any help from humans which is possible by using AI, ML, computer vision, neural network, etc., these are now being used to reduce the pollution caused by various forms of transportation. This technology takes the environment into complete surveillance with the help of sensors and IoT and then converts them into models and brings them inside ML which can improve the quality of the forecast report and when it is combined with AI it is possible to accurately predict the environmental situations of a particular environment in advance [18]. With those predicted data actions can be taken in advance to avoid risks.

In addition to the collection of sensor data, satellite data, traffic data, and other particular parameters AI and combined physics and chemistry model is used to blend the data from the sources and compare and combine it with the past data. This would result in the prediction of the future environment in advance. Smart machines can be designed to perform tasks with human intelligence. The intelligence of humans is simulated in machines by programming the machines and mimicking human actions.

4.5.1 STRENGTH OF ARTIFICIAL INTELLIGENCE (AI)

It is now possible to monitor real-time data using AI, which when harnessed gives actionable information to police officers who can anticipate and deter illegal activity. This improves passenger safety. In the same way, monitoring real-time data from autonomous vehicles (AVs) allows for just-in-time interventions to avoid collisions and ensure passenger safety. By proposing alternate routes and estimating wait times at traffic signals, smart roads that collect pedestrian and vehicular data are assisting in reducing traffic congestion and mitigating accidents.

After work, many people find it challenging to drive. In such situations, AVs may provide a rest stop for exhausted drivers, allowing them to nap in the car without endangering themselves or anyone on the road. Similarly, AVs driven by AI can be useful for people who find driving at night challenging. AI also aids in the detection and arrest of erring drivers, such as those who drink and drive or text while driving, making roads safer for all [10].

4.5.2 ARTIFICIAL INTELLIGENCE (AI) IN TRANSPORT POLLUTION

Using specialized techniques and algorithms, AI is addressing various transportation problems. The air quality index (AQI) is an efficient approach for pollution control. To protect citizens, the government gathers data and places restrictions on cars. Every year, millions of people die as a result of breathing heavily toxic air, while others suffer from serious health problems. ML techniques made a revolution in numerous areas in that air quality was classified and predicted by its classification and regression methods.

4.6 AIR QUALITY INDEX (AQI)

Knowledge about the quality of the air we breathe is important for every human being. It also helps both people and as well as the government to make various decisions. People can be aware of the current scenario of air quality by the resulting measurement of the AQI. It gives us a report of the quality of air as a routine. Air pollution is hitting the news in metropolitan cities like Delhi, every moment the quality of air increases and decreases [1]. When people breathe polluted air, it causes various serious health problems and even it leads to death. That's when this AQI comes into the picture, people can check the quality of the air they breathe instantly with the help of some mobile applications and websites.

The AQI is measured in the terms of two major resource concentrations of the pollutants for specific intervals of time [8]. The government alerts people about the current status of air pollution with the help of the AQI. If the AQI increases, so do the threats to public health. AQIs vary by country according to the emission of pollutants (Figure 4.4).

Air Quality	AQI Value
Safe	0–50
Average	51–100
Unsafe to breathe	101–150
Unsafe for healthy people	151–200
Highly dangerous	201+

FIGURE 4.4 Air quality index (AQI) table.

Air quality is the main attribute in the atmosphere. Determining the quality of the air is an expensive process in some techniques, developing countries can't afford the price of finding the quality of air. AQI is measured by ML techniques that are more cost-effective and resource-efficient [10]. The concentration of particulate matter in the air is predicted and classified using ML techniques. Several regression models have different and unique features to predict the quality of air. The predicted value should be accurate to take the appropriate measures.

4.6.1 LINEAR REGRESSION

Interaction between two continuous variables can be reviewed and revised with linear regression. This makes it possible to model the relationship between variables of scalar dependent with explanatory variables by a linear approach [7]. A simple linear regression has one descriptive inconsistent value; with more than one value, we have multiple linear regression. Linear regression data set has to be plotted in the graph with the x and y axis. Best fit line or regression line construction based on the least square method. Once the best fit line constructed process of minimizing vertical distance took place to minimize the error (Figure 4.5).

4.6.2 SUPPORT VECTOR REGRESSION

Support vector machine (SVM) is a supervised learning method that solves both classification and regression problems. Support Vector Regression carried out with the same feature of SVM. Support vector machine majorly solves classification problems, still helps to identify the robust and reliable values for regression problems [6].

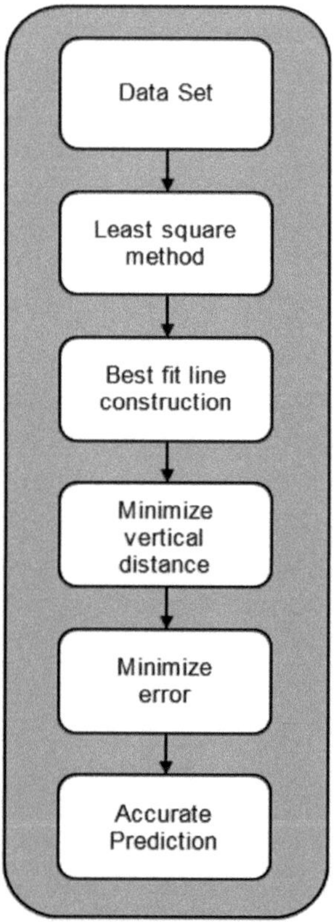

FIGURE 4.5 Working process of linear regression.

Support vector regression constructs a hyperplane with two decision boundaries for the given classes. Hyperplanes will be constructed by analyzing the class and providing the best place to solve the problem, and the positioning is completely dependent on extreme points and vectors. The output of air quality is going to be a continuous numeric variable such problems are solved efficiently using regression analysis. Non-linear separable data fed as input in finding the AQI and that data have to be transformed into high dimensional linear separable data using kernel tricks [4]. The goal of Support Vector Regression is to minimize the error in the prediction of the AQI (Figure 4.6).

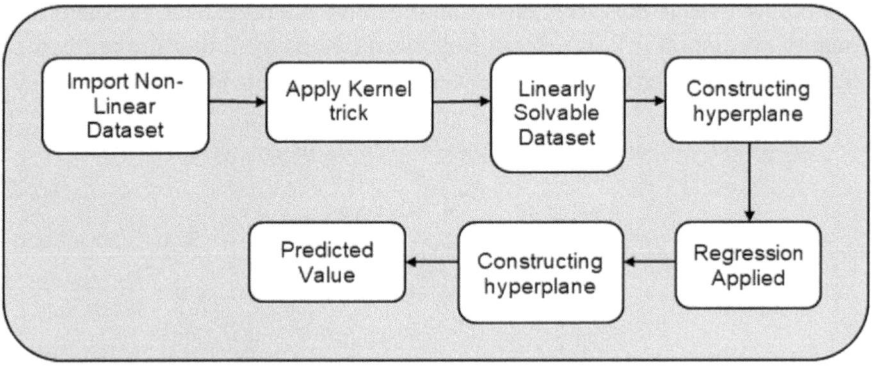

FIGURE 4.6 Working process of support vector regression.

4.6.3 RANDOM FOREST REGRESSION

Random forest algorithm is a bootstrap aggregation of the bagging method. The bagging method creates several base models with a sample data set and the conclusion drawn from those base models. Each base model gets trained on the given sample data. Bootstrap aggregation means data set trained separately in each base model and finally aggregates the result by combining each result of the base model using a voting classifier method.

Random forest regression collects the air quality data set as input data and considers the data to have m rows and n columns. Base learner models are designed to carry out the process of prediction by constructing decision trees. Each decision tree is supplied with a sample data set, sample data sets are drawn in rows and also with replacement [19]. Decision trees get trained parallelly with the row sampling with replacement and feature sampling as their input. Individual decision trees predict the air quality as the final output using the fed data set. Decision trees prediction has to be aggregated by analyzing all the decisions of each decision tree (Figure 4.7).

4.6.4 ADABOOST REGRESSOR

AdaBoost algorithm as its name says it boosts the accuracy of ML techniques so it is called a ML ensemble learning (EL) algorithm. Weak students are aided by their weights and require them to vote in the development of a combined final model. AdaBoost Regressor class to solve the regression problem of air quality prediction [6]. AdaBoost Regressor begins by fitting the regressor to the dataset and then adjusting the weights based on the error rate.

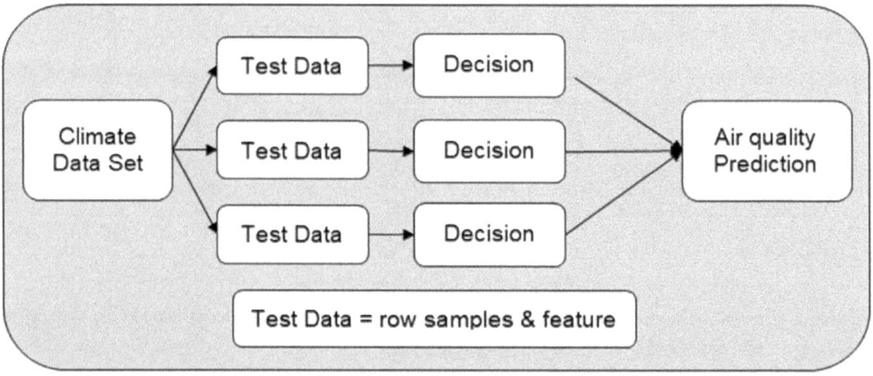

FIGURE 4.7 Working process of random forest regression.

Boosting assigns weights to classifiers based on their accuracy in predicting a random collection of input cases, and trains the base models one by one. As a result, the more accurate classifiers will contribute more to the final response. Weights are often assigned to each input object based on how difficult it is for all classifiers to predict the instance on average. The higher the weight, the more difficult it is to estimate the instance's ground truth, and therefore this object has a better chance of being the training subset in the next iteration (Figure 4.8).

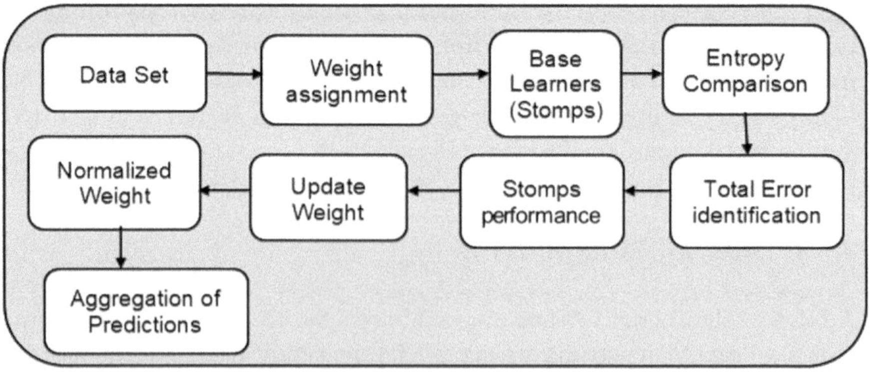

FIGURE 4.8 Working process of AdaBoost regression.

4.6.5 *RIDGE REGRESSION*

Ridge regression is a model tuning technique that can be used to analyze data with multicollinearity. L2 regularization is achieved using this approach. Where there is a problem with multicollinearity, least-squares are unbiased, and variances are high, the expected values are far from the actual values (Figure 4.9).

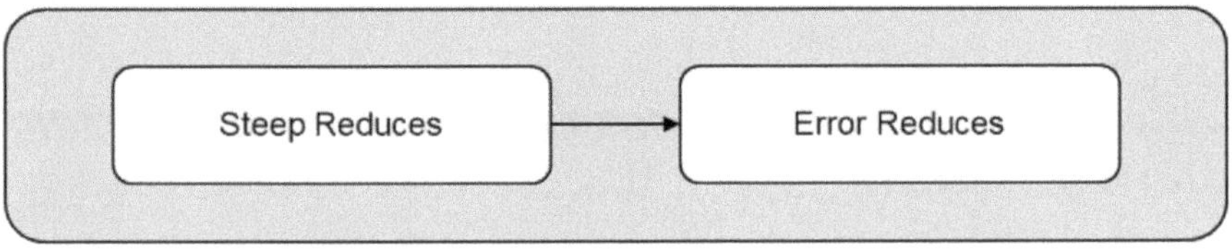

FIGURE 4.9 Working process of ridge regression.

4.7 AI-BASED TRAFFIC MONITORING SYSTEM

Technologies are getting smarter every day. This leads to finding useful solutions to many of the problems in society about 69% of the fuel consumed by the transportation sector but they are not consumed effectively. The vehicles standing in signals idle contribute CO_2 to the environment without any benefits. Around 60 billion gallons of fuel are wasted every year by vehicles idling in traffic [5]. There are many kinds of research going on around the world to design a traffic system that reduces idle time at the signs and during peak traffic times. This comes possible by combining and using AI and ML which will result in efficient traffic monitoring and regulating systems. Some of the tech leads use some of the technologies to detect the highly traffic-affected areas and indicate it to the people and also suggest an alternative route. Ex: google maps, AI, ML algorithms, vision sensors, radar, LIDAR, wireless communication, air quality sensors, temperature sensors, and audio sensors are used to analyze the current situation which helps to give priority to emergency vehicles like ambulances, police by directing the signals accordingly.

Vehicle tracking AI systems are tested in some of the countries by the researchers which detect the traffic at 95% accuracy [9]. It is possible to determine the speed of the vehicle, classification of the vehicles, and direction of the vehicle with the help of cameras that are placed above the traffic flow. About 400 traffic parameters can be determined by using the neural network at the intersection points [15]. The main aims of the smart traffic management system are to analyze real-time traffic circumstances to prioritize moderate traffic conditions, implementing an integrated E-bill payment system to improve conventional ticketing, warn those who violate traffic, to Provide public transit that is both secure and on time, to eliminate unwanted pollution due to traffic, Using GPS and GIS to implement smart traffic control systems at the intersections and road ends to provide accurate traffic guidance and better communication by optimizing road networks.

4.7.1 AI-BASED SOLUTIONS FOR SMART TRAFFIC MONITORING AND MANAGEMENT

A predetermined signaling scheme is being used in many cities, but it has many disadvantages. It does not consider the current traffic flow and works as defined earlier. But by using the smart traffic control system will analyze the real-time traffic and guide the signals accordingly. Here are some of the smart traffic management systems [11]:

- AI-enabled smart communication and traffic monitoring systems.
- Traffic signs, smart highways, public transit, smart parking, and a geospatial traffic guidance system are some of the technological elements that make up the smart traffic management system.
- Applications that automate conventional ticketing systems by offering a user-friendly interface.
- Vehicle tracking systems like GPS to trace the commutator location.
- Analyze data from the sensors mounted on the vehicle, road, and highways and inform the vehicle owners in case of any traffic violation.
- Mobile application that locates the free parking space at a particular location.
- AI integrated with GPS and sensors to identify less traffic and shortest distance routes to reach a particular destination.
- LIDAR, AI, ML techniques combined to detect the obstacles around the car which avoids many accidents.
- 3D visualization, RTK, and geographical data can be used to sense the situation of a particular location which allows the user to know the situation more accurately.
- Traffic queue length can be determined using vision sensors and GIS. And guiding the upcoming vehicle to a different route.

4.7.1.1 CONVOLUTIONAL NEURAL NETWORK (CNN)

Many efficient techniques are implemented to improve the quality of travel and also to reduce the environmental pollution caused by various transportation factors. All traffic signals are equipped with cameras, but they are used only for manual monitoring of traffic. But CNN implementation will help in automating the monitoring and detection process. Convolutional neural network (CNN) is a highly accurate algorithm that uses video data to classify the objects which can be used for the classification of road traffic. Compared to other classification algorithms, CNN has a minimized pre-processing. Evaluation is made after training the CNN. The trained model after the evaluation process is now ready to use and it can process the live streaming video and convert it into frames and show the traffic rating (Figure 4.10).

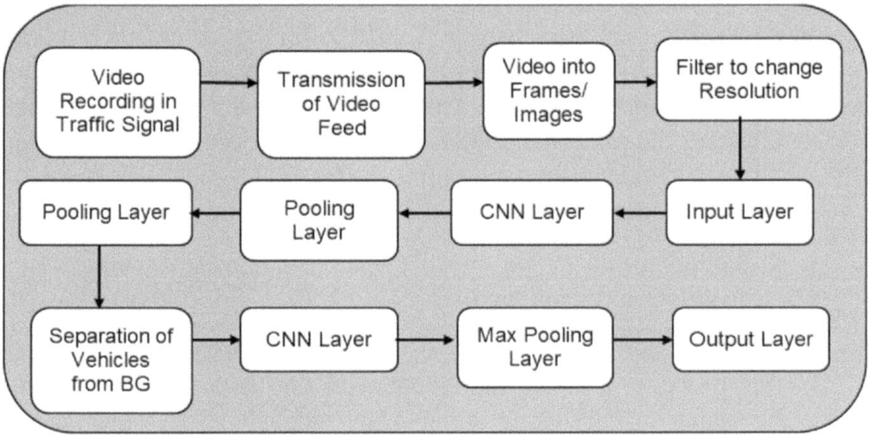

FIGURE 4.10 Working process of convolutional neural network.

CNN consists of neurons that have learnable weights and biases. A video is a combination of frames or images, and each image is a combination of pixels and each pixel RGB image holds some values which are called tensors that are used as the input for the neural network [13]. CNN is made up of multiple layers which are separated by supporting layers. Each layer is made up of neurons, and it is attached to neurons of the previous layer, and the connection goes on to the end of the layer. The classification is made by the end layer and it gives a rating from 0 to 10.

The pooling layer is the separation layer that acts as a filter and it modifies the resolution of each frame and output as well. The live video feed is taken from the traffic camera and it is converted into frames and each pixel in the images are analyzed by using CNN and removes the background and separates the vehicles only and after processing it gives the output as a rating ranging from 0–10 which is the classification of traffic level in different lanes by using this technique it is possible to identify the traffic levels in each lane and can guide the signals accordingly to reduce the traffic and also to reduce the pollution caused by the vehicles that are idle in traffic.

4.7.1.2 ARTIFICIAL NEURAL NETWORK

Various techniques and methods are used to manage traffic. Among the intelligent methods, Artificial Neural Network is one of the efficient methods. ANN combined with IoT, ML offers a completely reliable solution to the traffic congestion problems. There is also the involvement of some sensors and wireless networks. Traffic density and average speed of vehicles are the main parameters that are considered. Using this data, the Artificial Neural Network classifies the traffic level at each lane.

The entire process is divided into three main segments:

- Collection of data;
- Transmission of data; and
- Processing of data.

In this process, ANN is implemented in the third segment to determine the status of each congestion. ANN helps in processing the data intelligently [14]. Artificial neural networks (ANNs) consist of multiple neurons that are connected by weights. The input layer, an output layer, and a hidden layer are the three main layers of the neural network. The traffic data such as vehicle speed and traffic density values are fed as input to the input layer, and no processing is done in this layer. It transfers the values as numbers to the hidden layers which consist of 5 neurons in this condition, and it does all the processing tasks and transfers the output as numbers to the output layer. The output layer consists of the average density of vehicles in a particular lane and the average speed of vehicles in traffic. Before all these the ANNs should be trained with the pre-existing data set. In real-time the traffic data is taken from the sensors and sent to the processing unit by the transmission unit and processing is done using the ANN method and the output is displayed [12]. And with the output values, the traffic can be guided accordingly. Queue length can be reduced and idle time of vehicles can be reduced and pollution produced by the vehicles can be controlled.

4.8 CONCLUSION

Pollution spoils our environment in multiple ways. In this chapter, pollution causes and pollutants which are polluting our environment are discussed in detail. Pollution damages agricultural land, air, and water which are essential for living organisms. Air pollution spreads quickly in our environment and is also the reason for multiple health problems in humans, animals, and even also for plants. The WHO released a shocking survey that 9 out of 10 humans breathing polluted air which induced many scientists around the world to work on reducing the impact of air pollution. Inventions like the automated car of Elon Musk to AQI prediction by TensorFlow by college students of Delhi are showing their responsibility to protect our environment. An increase in the number of vehicles is the major reason for transport pollution. Modern techniques are opening their hands to address the problem of transport pollution. AI predicts the quality of air, classifies the pollutants, and provides the best solutions for traffic congestion, AVs, and many more inventions that laid the groundwork for the betterment of the environment. We believe and work for a pollution-free environment in the future along with the great technology AI.

KEYWORDS

- **air quality index**
- **air quality micro-climate monitoring system**
- **artificial neural networks**
- **hydrocarbons**
- **International Maritime Organization**
- **polycyclic aromatic hydrocarbons**

REFERENCES

1. Ma, L., Graham, D. J., & Stettler, M. E., (2021). Air quality impacts of new public transport provision: A causal analysis of the jubilee line extension in London. *Atmospheric Environment, 245,* 118025. https://doi.org/10.1016/j.atmosenv.2020.118025.
2. Liang, Y. C., Maimury, Y., Chen, A. H. L., & Juarez, J. R. C., (2020). Machine learning-based prediction of air quality. *Applied Sciences, 10*(24), 9151. https://doi.org/10.3390/app10249151.
3. Jiang, Y., Xing, J., Wang, S., Chang, X., Liu, S., Shi, A., Liu, B., & Sahu, S. K., (2020). Understand the local and regional contributions on air pollution from the view of human health impacts. *Frontiers of Environmental Science & Engineering, 15*(5). https://doi.org/10.1007/s11783-020-1382-2.
4. Leong, W., Kelani, R., & Ahmad, Z., (2020). Prediction of air pollution index (API) using support vector machine (SVM). *Journal of Environmental Chemical Engineering, 8*(3), 103208. https://doi.org/10.1016/j.jece.2019.103208.
5. Weerasekera, R., Sridharan, M., & Ranjitkar, P., (2020). Implications of spatiotemporal data aggregation on short-term traffic prediction using machine learning algorithms. *Journal of Advanced Transportation, 2020,* 1–21. https://doi.org/10.1155/2020/7057519.
6. Halsana, S., (2020). Air quality prediction model using supervised machine learning algorithms. *International Journal of Scientific Research in Computer Science, Engineering and Information Technology*, 190–201. https://doi.org/10.32628/cseit206435.
7. Madhuri, V., Gunjal, S., & Kamalapurkar, S., (2020). Air pollution prediction using machine learning supervised learning approach. *International Journal of Scientific & Technology Research,* 118–123. http://www.ijstr.org/final-print/apr2020/Air-Pollution-Prediction-Using-Machine-Learning-Supervised-Learning-Approach.pdf (accessed on 1 January 2023).
8. Aarthi, A., Gayathri, P., Gomathi, N. R., Kalaiselvi, S., & Gomathi, V., (2020). Air quality prediction through regression model. *International Journal of Scientific & Technology Research*, 923–928. http://www.ijstr.org/final-print/mar2020/Air-Quality-Prediction-Through-Regression-Model.pdf (accessed on 1 January 2023).
9. Atmakuri, K. C., & Prasad, K. V., (2020). A comparative study on prediction of Indian air quality index using machine learning algorithms. *Journal of Critical Reviews*, 41–46. https://doi.org/10.31838/jcr.07.13.05.
10. Ly, H. B., Le, L. M., Phi, L. V., Phan, V. H., Tran, V. Q., Pham, B. T., Le, T. T., & Derrible, S., (2019). Development of an AI model to measure traffic air pollution from multisensor and weather data. *Sensors, 19*(22), 4941. https://doi.org/10.3390/s19224941.
11. Dhanaraj, R. K., Krishnasamy, L., Geman, O., & Izdrui, D. R., (2021). Black hole and sink hole attack detection in wireless body area networks. *Computers, Materials & Continua, 68*(2), 1949–1965.
12. Abduljabbar, R., Dia, H., Liyanage, S., & Bagloee, S. A., (2019). Applications of artificial intelligence in transport: An overview. *Sustainability, 11*(1), 189. https://doi.org/10.3390/su11010189.
13. Mashik, A., (2019). *Machine Learning Algorithms in Air Quality Modeling.* https://www.gjesm.net/article_35967_bd391739196a2224b2ff75a4fba6da20.pdf (accessed on 1 January 2023).

14. Dhingra, G., Supreeth, S., Neha, K., R., Amruthashree, R. V., & Eshitha, D., (2019). *Traffic Management Using Convolution Neural Network,* 146–149. https://www.ijeat.org/wp-content/uploads/papers/v8i5S/E10310585S19.pdf (accessed on 1 January 2023).
15. Mondal, A., & Rehena, Z., (2019). *Intelligent Traffic Congestion Classification System Using Artificial Neural Network,* 110–116. https://doi.org/10.1145/3308560.3317053.
16. Tian, Y., Zhu, Q., Lai, K. H., & Venus, L. Y., (2014). Analysis of greenhouse gas emissions of freight transport sector in China. *Journal of Transport Geography, 40,* 43–52. https://doi.org/10.1016/j.jtrangeo.2014.05.003.
17. Azeez, O., Pradhan, B., Shafri, H., Shukla, N., Lee, C. W., & Rizeei, H., (2019). Modeling of CO emissions from traffic vehicles using artificial neural networks. *Applied Sciences, 9*(2), 313. https://doi.org/10.3390/app9020313.
18. Aditya, C. R., Chandana R. Deshmukh, Nayana, D. K., Praveen Gandhi Vidyavastu & Astu, V., (2018). Detection and prediction of air pollution using machine learning models. *International Journal of Engineering Trends and Technology, 59*(4), 204–207. https://doi.org/10.14445/22315381/ijett-v59p238.
19. Bai, L., Wang, J., Ma, X., & Lu, H., (2018). Air pollution forecasts: An overview. *International Journal of Environmental Research and Public Health, 15*(4), 780. https://doi.org/10.3390/ijerph15040780.
20. Veljanovska, K., & Dimoski, A., (2018). Air quality index prediction using simple machine learning algorithms. *International Journal of Emerging Trends & Technology in Computer Science, 7*(1), 25–30. https://www.ijettcs.org/Volume7Issue1/IJETTCS-2018-02-06-15.pdf (accessed on 1 January 2023).
21. Kleine, D. J., Zalakeviciute, R., Gonzalez, M., & Rybarczyk, Y., (2017). Modeling PM2.5 urban pollution using machine learning and selected meteorological parameters. *Journal of Electrical and Computer Engineering, 2017,* 1–14. https://doi.org/10.1155/2017/5106045.

CHAPTER 5

Neural Network Approaches for Transportation

N. R. GAYATHIRI[1], S. NITHYAPRIYA[1], T. MALATHI[1], A. KODIESWARI[2], and K. R. SABARMATHI[2]

[1]*Department of Artificial Intelligence and Data Science, Bannari Amman Institute of Technology, Erode, Tamil Nadu, India*

[2]*Department of Information and Technology, Bannari Amman Institute of Technology, Erode, Tamil Nadu, India*

ABSTRACT

In the current generation, neural networks are used in various sectors like education, finance, speech recognition, face detection, aerospace, automotive, transport, stock market prediction, transportation, etc. Transportation system plays a vital role in economic development. Sustainable and acceptable artificial neural networks (ANNs) techniques are used to make the transportation system intelligent. We use neural network approaches for various transportation issues like traffic forecasting, route planning, vehicle number plate detection, parking space identification, vehicle density, travel time, autonomous vehicle, vehicle location, vehicle speed, driver behavior, traffic management, etc. The neural networks algorithm helps to make the decision like human beings. So, this chapter presents intelligence in the transportation system through the efficient implementation of neural networks. Neural Network simplified design configuration is the added advantage for making the intelligent transportation system.

Artificial Intelligence for Future Intelligent Transportation: Smarter and Greener Infrastructure Design. Rajesh Kumar Dhanaraj, Nilayam Kumar Kamila, Subhendu Kumar Pani, Balamurugan Balusamy, & Vani Rajasekar (Eds.)

5.1 INTRODUCTION

The neural network is a computer model with interrelated nodes that have characteristics like the human neurons to process the information. Neural networks can pick out the hidden styles and correlations in raw facts, group, and classify data, and as time go through, it learns and improves with the assistance of algorithms [8].

Basic functions of neural networks:

1. **Classification:** Arrange patterns or datasets into predetermined classes.
2. **Prediction:** Generate the anticipated output from a given input.
3. **Clustering:** Identify a unique characteristic of the data and categorize them without any prior knowledge of the data.
4. **Associating:** Neural networks are trained to remember patterns and models. When we show an unknown version of a model, the network associates it with the most comparable version of its memory and returns to the last [14].

5.2 HOW DO NEURAL NETWORKS OPERATE?

Neural networks are made up of a large number of artificial neurons, which are called nodes that are organized in a sequence of layers. The input layer is the first layer that accepts inputs in multiple different formats supplied by the programmer. There is one more layer called the hidden layer present between the input and output layer. The hidden layer is used to perform all computations in order to find hidden features and patterns. Using the hidden layer, the input passes through a series of transformations which ultimately leads to an output which is displayed using the output layer.

Input data from the input layer moves into hidden layers for processing. In the processing stage, every input data is assigned with appropriate weights. Each input data is multiplied by its weights. Then, summing of input-weight takes place. Bias is the additional parameter added along with a weighted sum of input to adjust the output. Bias added weighted sum of input is passed to the activation function. The activation function decides the output that the node is going to generate based on the input data. Finally, the output is retrieved from the output layer. If the expected output is not generated, back propagation has to be done.

Neural networks are an example of a supervised learning algorithm and aim to approximate the function represented by the data. This is accomplished by calculating the error between projected and expected outputs during the training phase and minimizing it (Figure 5.1).

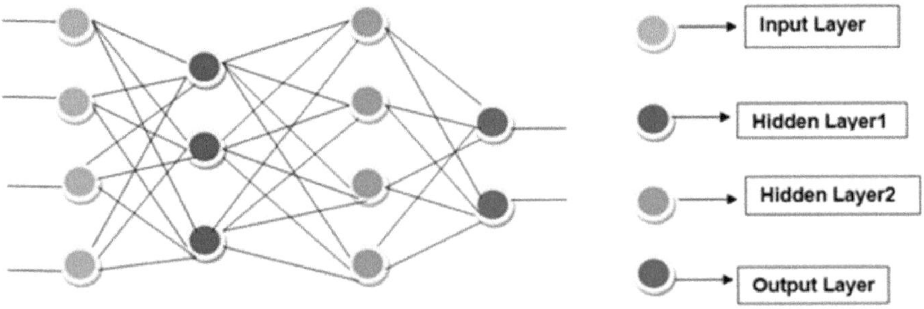

FIGURE 5.1 Workflow of neural networks.

5.3 WORKING PROCESS OF SINGLE NODE

Here the product inputs (X_1, X_2, X_3, …, X_n) in the input layer and weights (W_1, W_2, W_3, …, W_n) in the hidden layer are summed with bias (b) and then acted on by an activation function f(y). In a neural network, the activation function is the most important aspect in determining whether or not a neuron in the network will be activated and transported to the next layer. This essentially implies that it will determine whether or not the input to the neural network is significant to the prediction process. The activation functions are also known as threshold or transformation. Due to their differentiable nature, it aids in the normalization of the output between 0 and 1 or –1 to 1 in the back propagation process (Figure 5.2).

The different types of activation functions:

- Binary step;
- Leaky ReLU;
- Softsign;
- SoftMax;
- Swish;
- ReLU;
- Parameterized ReLU;
- Tanh;

- Exponential linear unit;
- Sigmoid;
- Linear.

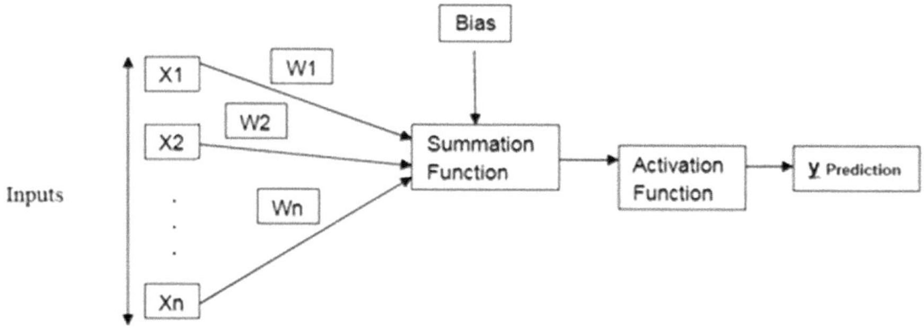

FIGURE 5.2 Workflow of single node in neural networks.

5.4 OPTIMIZATION ALGORITHMS

The approach used to carry out the learning process is known as the training (or learning) strategy. In order to produce the best possible outcomes, the neural network is trained using the training technique. The manner in which the parameters of the neural network are adjusted determines the type of training. Neural network can learn from its past experience. Learning process is carried out by the Optimization algorithm. Optimization algorithms are of different types based on memory, processing speed and other characteristics:

- Gradient descent;
- Adam;
- SGD with momentum;
- Adaptive gradient (AdaGrad);
- Newton method;
- AdaDelta;
- Quasi-Newton method;
- Conjugate gradient;
- Levenberg-Marquardt algorithm;
- Nesterov accelerated gradient (NAG);
- Stochastic gradient descent (SGD);
- Mini batch stochastic gradient descent (MB-SGD).

5.5 INTELLIGENT TRANSPORT SYSTEM APPLICATIONS

Various types of applications of neural networks in the Intelligent Transportation System are presented in the chapter [2]:

- Traffic light recognition;
- Traffic sign recognition;
- Traffic forecasting and control;
- Evaluation of traffic parameters;
- Traffic signal timing detection;
- Vehicle classification;
- Driver behavior recognition;
- Road damage detection;
- Lane/turn recognition;
- Prediction/understanding traffic scene;
- Vehicle attribute detection;
- Maintenance of transport infrastructure;
- Autonomous vehicles;
- Vehicle parking;
- Predict bus arrival time;
- Travel time prediction;
- Detecting vehicle location and estimation of the vehicle speed on the road;
- License plate detection;
- Pedestrian detection.

5.6 CLASSIFICATION OF NEURAL NETWORKS

Based on the mathematical model, principles, and parameters, neural networks are classified. Each neural network has different learning processes and also has its own advantages and disadvantages. The most common types of neural networks in use today and for what they are used are discussed.

5.6.1 FEED-FORWARD NEURAL NETWORK

In this type of neural network, data travels from the input layer to the output layer through hidden layers (if any) only in the forward direction.

The data passes only in the forward and never backward direction, so cycles are not formed in the network. Feed-forward neural network does not remember the past as there is no input memory. So, it does not predict the future. It remembers only the training.

5.6.1.1 APPLICATIONS: RECOGNITION OF FACE AND COMPUTER VISION (Figures 5.3 and 5.4)

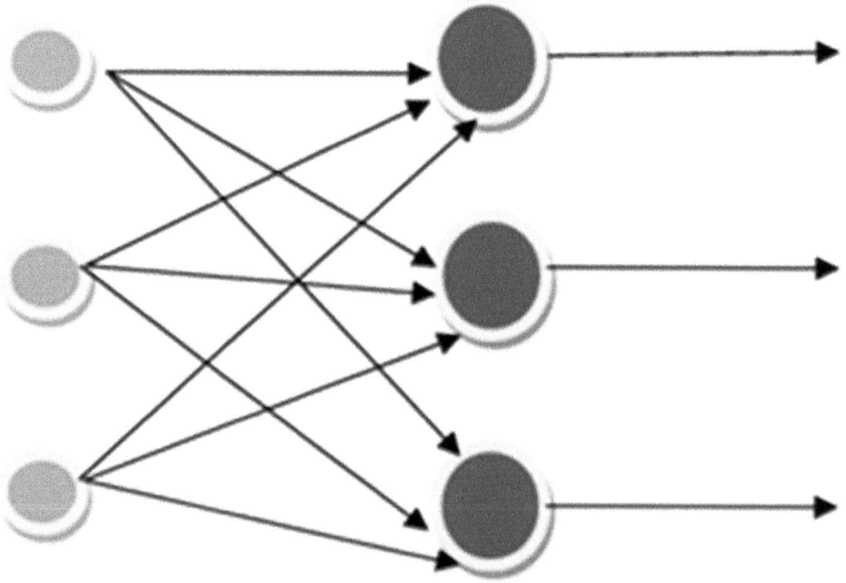

FIGURE 5.3 Workflow of single node in neural networks.

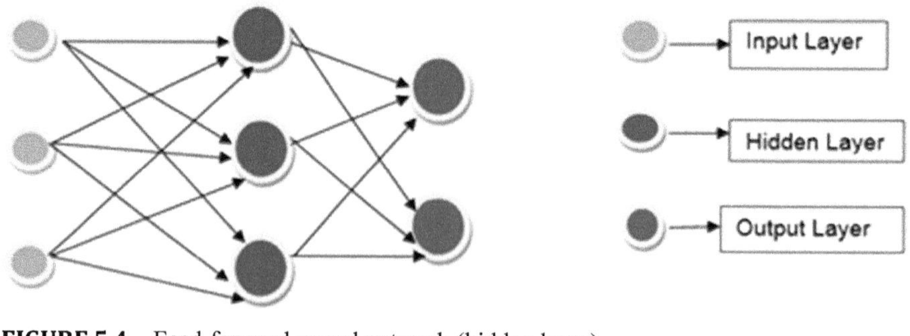

FIGURE 5.4 Feed-forward neural network (hidden layer).

5.6.1.2 SINGLE LAYER PERCEPTRON

If the hidden layer does not exist in the feed forward neural network, then it is called single layer perceptron. Input data is assigned with random weights because there is no past knowledge. The product of input-weight is summed up. Threshold value is fixed for each perceptron. The perceptron is activated only if the summation value of input weight is above the threshold value. If the received output is not the same as the predicted output, then weights assigned to the input data are changed to get the desired output (Figure 5.5).

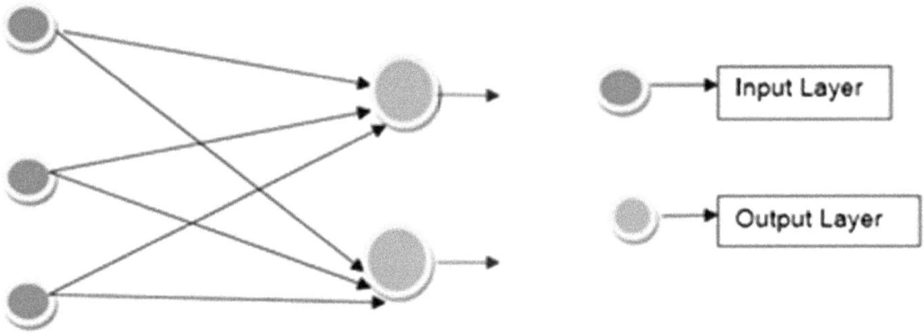

FIGURE 5.5 Single layer perceptron.

5.6.1.3 MULTILAYER PERCEPTRON

If an arbitrary number of hidden layers exist in the Feed Forward Neural Network, then it is called a Multi-Layer Perceptron. Input data from the input layer passes through many hidden layers to process the input data. From the hidden layer processed data goes to the output layer to display the output. In this process, hidden layers are the computational engine. Multilayer perceptron is trained to relate the input-output correlation. To get the predicted output parameters, bias, and weights are adjusted. Adjustment is done by back propagation method (Figure 5.6).

5.6.2 CONVOLUTIONAL NEURAL NETWORK (CNN)

Another type of neural network is convolutional neural network (CNN) which is formed by different layers. Each layer is explained below (Figure 5.7):

1. **Input Layer:** It is to feed the input (image) data into the model for processing.
2. **Convolutional Layer:** This mines the different features from the image data.
3. **Pooling Layer:** It is used to decrease the computational factors and it is present between the successive convolution layers.
4. **ReLU Layer:** It is present for removing negative pixels by zero (non-linearity).
5. **Fully Connected Layer:** It is used to connect the neurons present in one layer to the next layer by a multi-layer perceptron. Every neuron learns from the past experience and that learning helps to predict the output.

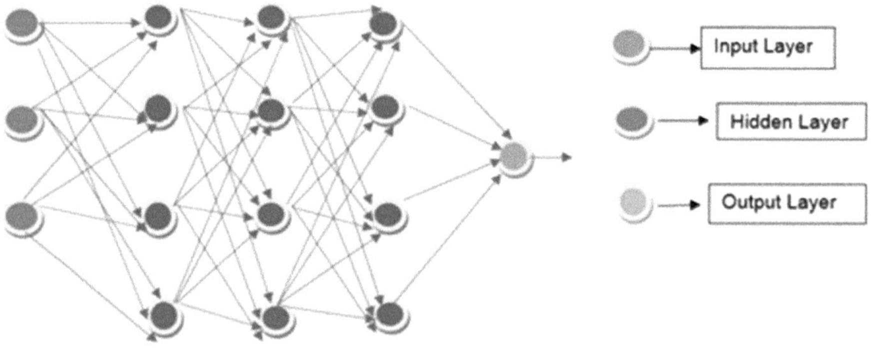

FIGURE 5.6 Multi-layer perceptron.

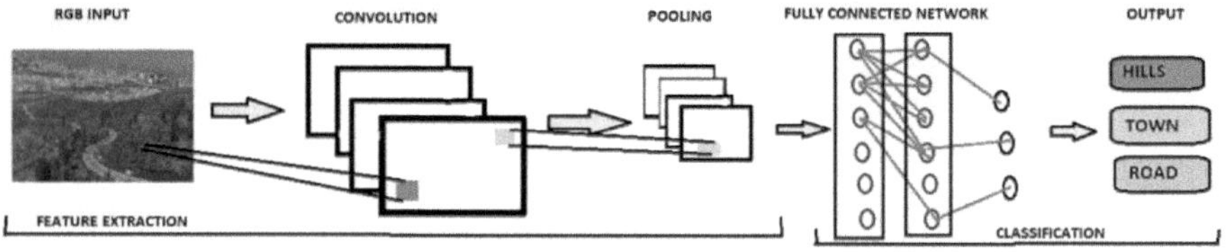

FIGURE 5.7 Convolutional neural network.

The foremost application area of CNN is image recognition and image classification. If the desired output is not achieved, then do back propagation by adjusting the weight and bias to get the predicted output. Less preprocessing is the main advantage of convolution neural networks. Less preprocessing is achieved by kernels (set of filters).

5.6.3 RECURRENT NEURAL NETWORK (RNN)

Each node present in the recurrent neural network (RNN) acts as a memory cell, to store the output of the previous layer. Now, the input for the next layer will be the previous layer output. RNN uses the above method to predict the outcome of a layer. This type of neural network uses front propagation first and stores all processed information for future use. If the predicted output is not the same as the received output, then we want to make a neural network to self-learn to correct the mistake by back propagation method. The main application of RNNs is text to speech conversion (Figure 5.8).

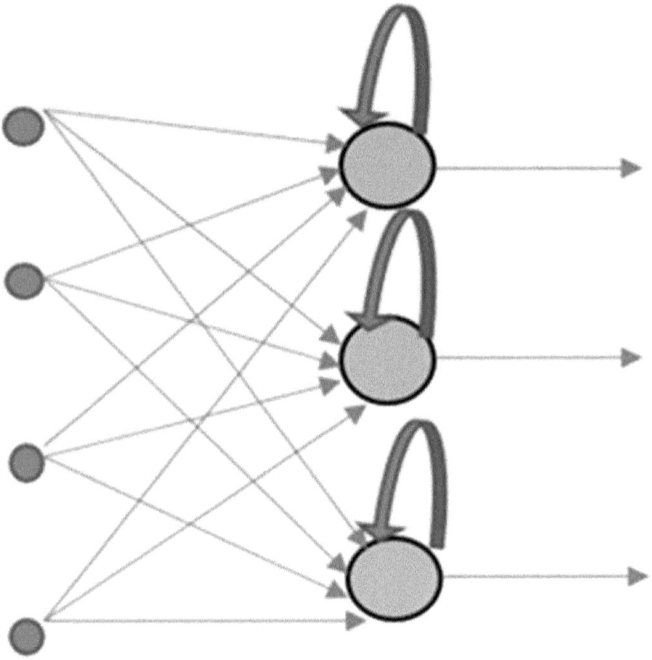

FIGURE 5.8 Recurrent neural network (RNN).

5.6.4 GRAPH NEURAL NETWORK (GNN)

Graph neural networks (GNNs) resemble the graph data structure. If two nodes are interrelated, then those two nodes should maintain its neighbor nodes information (embeddings). This is achieved by GNNs by connecting the nodes through edges. This neighborhood information is used to solve many problems like node and edge prediction. The GNN is widely used in our day-to-day applications. Its importance is remarkable one. Social networks, Google maps, Chemical compound linking, recommender systems, and route map in transportation are quite a few applications where GNNs are used.

5.6.5 WORKING OF GRAPH NEURAL NETWORKS (GNNS)

The aim of GNN is that each node should learn about the other nodes in the network. In the above figure, Node A and Node C are the neighbor nodes for Node B. Node B collects Node A and Node C information. After that, Node B adds its own information and passes whole embeddings (A, B, C) to the main Node A. The above process continues until all the nodes in the network know about other nodes. The same process is repeated for Node C and Node D. Now main node A contains whole graph embeddings (Figure 5.9).

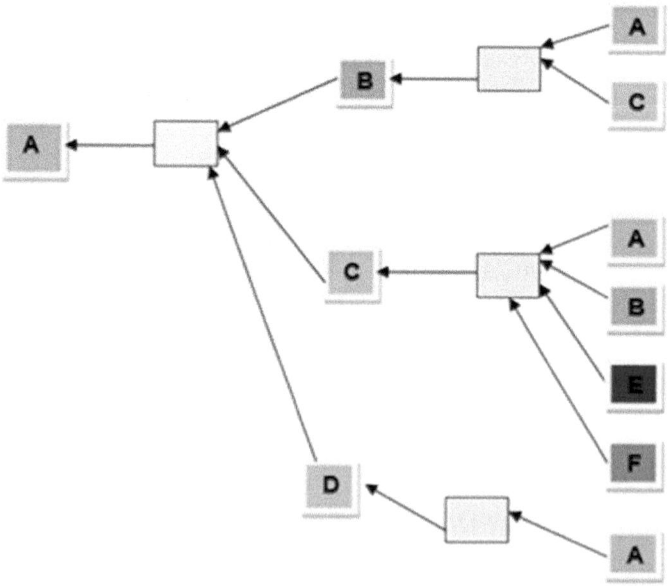

FIGURE 5.9 Graph neural network (GNN).

5.7 CONVOLUTIONAL NEURAL NETWORK (CNN) IN TRANSPORTATION

5.7.1 TRAFFIC CHARACTERISTICS PREDICTION

Traffic is the major problem in transportation. Using the intelligence of transportation, traffic can be reduced by predicting the traffic characteristics. This prediction will help the drivers to choose routes more efficiently. Vehicle speed, traffic density, flow, and conditions are the main traffic characteristics [1].

The major problem in the traffic prediction is irregular traffic density. We cannot predict the time accurately for heavy traffic. It varies from time to time. But irregular traffic can be predicted with high accuracy using deep learning methods. Instead of using parametric and statistical methods, recently non-parametric methods give higher prediction for traffic flow.

5.7.2 TRAFFIC INCIDENTS INFERENCE

Accidents are quite common in some locations due to heavy traffic. Heavy traffic results in traffic incident risk. We want to predict the traffic incident risks based on the traffic parameters in the location. Driver Speed, Weather, Vehicle Flow, Human Mobility, Geographical position of the location, Traffic flow are some of the parameters considered for traffic incidents risk. But the above parameters will differ from place to place and time to time for the same location. All the model steps have to be stored in the data pool to predict incident severity in the future.

5.7.3 VEHICLE IDENTIFICATION

Automatic vehicle identification plays a vital role in toll gate and vehicle parking. Vehicles are identified with the number plates which are unique for each vehicle.

Number plate recognition becomes difficult due to the parameters like Low quality camera, less lightening and blurriness due to vehicle movement and number plates are hidden by other vehicles. To overcome the above drawback, a deep learning visual model is used. Reading Vehicle number plates are automatically recognized by the combination of support vector machine (SVM) and CNN. In case of, similar number plate Siamese neural network is used [1]. CNNs play a vital role in recognizing license plate, vehicle color, and vehicle type classification.

5.7.4 TRAFFIC SIGNAL TIMING

Automatic traffic signal change is very important to reduce traffic and accidents. Based on the time, traffic signal lights are changed. In order to reduce the risk, traffic signal lights have to be changed based on the traffic density. This can be done by deep CNNs. In the first stage CNN extracts the raw live data like vehicle current position, speed of the vehicles and other objects mobility. In the second stage, raw data is given to the deep reinforcement learning (RL) network. This reinforcement network is responsible for predicting the automatic change of traffic signal lights (red, green, yellow), waiting time and safe left turn. In this way traffic signal light is changed according to the density of the vehicle not based on the time [1].

5.7.5 RIDE SHARING AND PUBLIC TRANSPORTATION

Frequent Public transportation has to be provided to the places where passenger demand is more. Similarly, when there is no passenger demand, the frequency of public transportation has to be reduced. Vehicle routing has to take place as per the passenger flow. CNN is used to predict the travel mode, passenger flow, passenger demand prediction [1].

5.7.6 VISUAL RECOGNITION TASKS

Autonomous vehicles (AVs) are the booming technology in transportation. Driverless vehicles depend mainly on visual recognition. Camera image-based systems along with the sensors is used for object detection like obstacle detection [1].

5.7.7 RATE THE ROAD LANE-BASED ON INCOMING TRAFFIC

CNN is used to rate the road lanes based on the incoming traffic. The traffic signal lights are changed based on the road lane classification rate (heavy, moderate, less traffic), which is more useful for traffic management. The road lane videos which is taken from 360° camera is given as input in the form of video frames to the CNN. A large amount of video frames is captured by traffic cameras and pass the video frames as training data to the neural network for better prediction. The output of CNN is analyzed with the traffic signals and the road lanes with high traffic rate value is allotted with more traffic signal light time [12].

5.7.8 LICENSE PLATE RECOGNITION

Identification of License plate is difficult task due to various shape, font, color, and light effects. The alphabets and numbers in the license plate is segmented using a clustering algorithm. Then, CNN is used to find the alphabets and numbers in the license plate [11].

5.8 OTHER APPLICATIONS OF CONVOLUTIONAL NEURAL NETWORKS (CNNS)

Obstacle detection, vehicle number plate identification, vehicle color and model detection, traffic sign recognition and auto changing, traffic light recognition, free lane detection, labeling the scene around place, traffic scene segmentation, road crack detection are some applications where CNNs is used for intelligence transportation.

5.9 CONVOLUTIONAL NEURAL NETWORKS (CNNS) IN SELF-DRIVING CARS

Decision process is autonomous in self-driving vehicles. The concept behind the automatic decision-making process is deep learning algorithms. These algorithms process the data, which is collected from different sensors like RADAR, GPS, LiDAR. Among the deep learning algorithms, CNNs are the most important algorithm for image recognition and classification. Self-driving car uses CNNs for recognition and classification of the various parts of the road and also for taking the correct decision [3]. The main components of driverless cars are explained in subsections.

5.9.1 PERCEPTION

To drive the car safely, the driver has to recognize and classify the things around the car environment. Recognition and classification of things takes place through the driver's eyes. In the driverless car, cameras, and sensors are used to recognize and classify the objects like pedestrians, parking palace, traffic signal lights, road signs, road turns and parallel vehicles. In order to avoid accidents, the exact distance between driverless cars and the surrounding objects are to be calculated automatically. Based on the distance calculation, the car has to slow down automatically.

For the self-driving cars, sensors like LiDAR, Camera, and RADAR act as human eyes for good perception. Figure 5.10 shows the AI powered autonomous car perception-planning-action to take the driving decision.

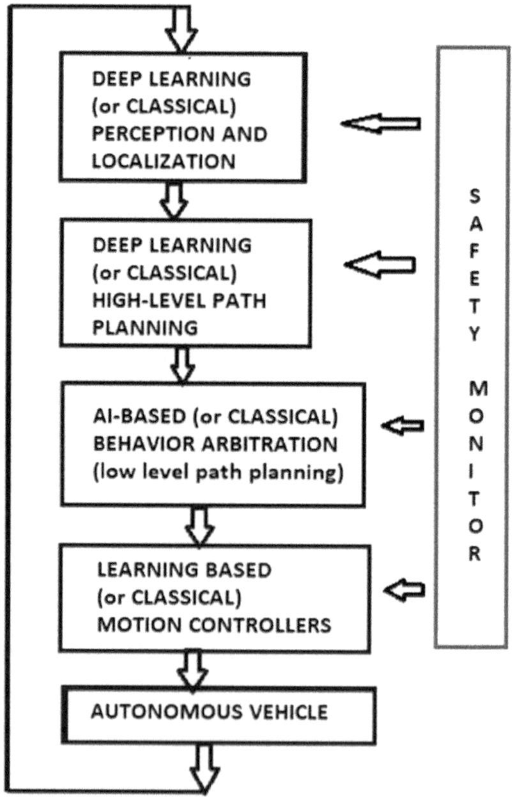

FIGURE 5.10 AI powered autonomous car perception-planning-action.

5.9.2 CAMERA

The high-quality camera acts as the human eye for driver less cars. Camera is used for segmentation, classification, and localization. High resolution Camera captures the surrounding environment images of the car more accurately [5].

5.9.3 LIDAR

The abbreviation for LiDAR is light detection and ranging. From the Lidar Sensor, laser light is fired. The distance between driverless cars and other objects is measured by how far the laser beam is reflected by the surrounding objects. Camera captures the surrounding images. Combination of camera and LiDAR sensor gives the detailed information of surrounding objects around the car. This detailed information is given as input to the deep neural networks for taking the safest decision. This is very useful in multi-exit driving conditions.

5.9.4 RADARS

In the LiDAR sensor, the distance between driverless cars and other objects is measured by laser beam. But Radio detection and ranging (RADAR) uses radio signals to calculate the distance between driverless cars and other objects. Even high-quality cameras cannot capture some objects, but uncaptured objects are detected by radio signals [5].

5.9.5 LOCALIZATION

Location and inclination of the car changes as the vehicle moves. The Localization algorithm uses consecutive frames of the video for calculating the Location and inclination of the car called visual odometry (VO). Simultaneous localization and mapping (SLAM) algorithm are used to find the location and inclination of close vehicle by referring to the previous frame.

5.9.6 PREDICTION

The high-quality camera in the self-driving car gives the 360-view image. All the camera images, Data from the LiDAR and RADAR sensors are given as input data to the deep neural network and then it processes the input data for decision making. Processed Data in different formats are used for object detection, classification, localization, and prediction [5].

Due to the more hidden layers in deep learning networks, data from different sources are processed accurately for making the correct decisions. Deep layers help the car to take the accurate inference for changing the route with actions like speed slowly, halt, brake, and taking the different route. Deep learning is used for computer vision and perception [5].

5.9.7 DECISION-MAKING

Driverless cars should take the decision as the human being even in unexpected situations. Driverless cars should take the accurate decision like slowing down the car, changing the route and turning the car, when the person is crossing the road all of a sudden. Sensors are used to collect the data from the different sources to make the correct decision. Position of the car and car movements are decided by deep learning algorithm [5].

Deep RL is used for the decision-making process. The Markov decision process is used for future decisions and route planning. Networks learn from the past information to predict the future behavior. Route planning, Behavior Arbitration, Motion Planning and Vehicle Control are the different components used in self-driving cars for taking decisions [5].

5.9.8 ROUTE PLANNING

The destination is fed into the car by the passenger. The driverless car should take the best route with less traffic, good road conditions, and less time. If the obstacle is present, the car has to take the decision by stopping the car or changing the car inclination.

5.9.9 BEHAVIOR ARBITRATION

Markov decision process decides the route planning. But the route plan has to be changed if there is uncertainty like sudden vehicle cross or natural calamities [5].

5.9.10 MOTION PLANNING

Final destination is reached by the passenger through the Motion planning activities like route changing and speed change. Motion planning is dynamically changed based on the current surroundings of the car [5]. Self-driving car architecture is given in Figure 5.11.

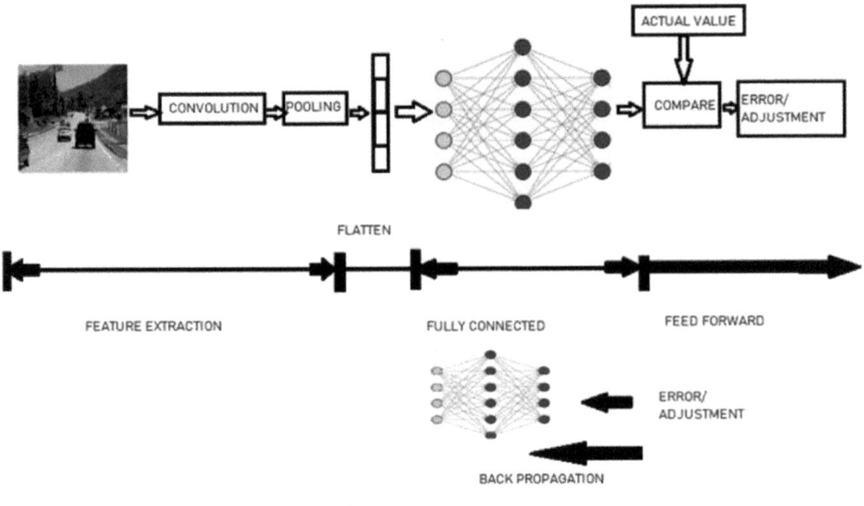

FIGURE 5.11 Self-driving car architecture.

5.9.11 VEHICLE CONTROL

The passenger has to reach the final destination with utmost safety. When the car speed is increased or decreased suddenly, the vehicle speed should be in control ensuring the surrounding safety.

CNN extracts the hidden features/pattern from the data collected from the high-quality camera and sensors. The neural network goes deeper by adding more hidden layers to extract the complex features like the shape of the tire, and surrounding objects. Because of the above reason, CNN acts as the main algorithm [4]. Nvidia, Tesla, Google Waymo are the big giant companies involved in making self-driving cars using CNNs.

5.10 GRAPH NEURAL NETWORK (GNN) FOR TRAFFIC FORECASTING

Graph attention and graph convolutional networks are the important types of GNNs used for traffic forecasting problems. Speed, traffic flow, demand, and passenger flow forecasting are the major traffic forecasting problems [6].

GNNs capture spatial dependency and temporal dependency (road condition changes) for traffic forecasting problems. Let us represent a road network as a graph. Each road segment and sensor for the road is taken as a node and road connection is considered as edge. Best results are produced by GNNs in traffic forecasting problems [6].

When a person is planning for the tour, route selection and traveling time estimation are predicted by using GNNs. Companies like Google uses GNNs for traffic forecasting, route planning and traveling time estimation.

The road image is collected through cameras with many objects like vehicles, human, shop, animals, etc. This road image is passed to GNNs model to describe the relationship between the objects. Suppose two vehicles are close to each other we calculate the distance between them and pass the traffic flow forecasting to drivers to avoid accidents. We can also group the similar type of objects and estimate the total vehicles of the same type.

The traffic information is processed to divide the road networks into Super-segments. The Super-segments consists of a set of road segments. Each road segments have specific length and speed. Each neural network model is trained for each Super-segments vary in the length. So, we need separate trained neural network model for each super segments to handle the road infra-structure challenge [7].

Google map usage of GNN for determining optimal route and travel time is given in Figure 5.12 [7].

5.10.1 TRAJECTORY FORECASTING

In order to predict the trajectory of moving objects based on the past trajectory. Trajectory forecasting is vital in self-driving vehicles to take the decision. Moving agent's trajectories can be forecasted by GNN [15].

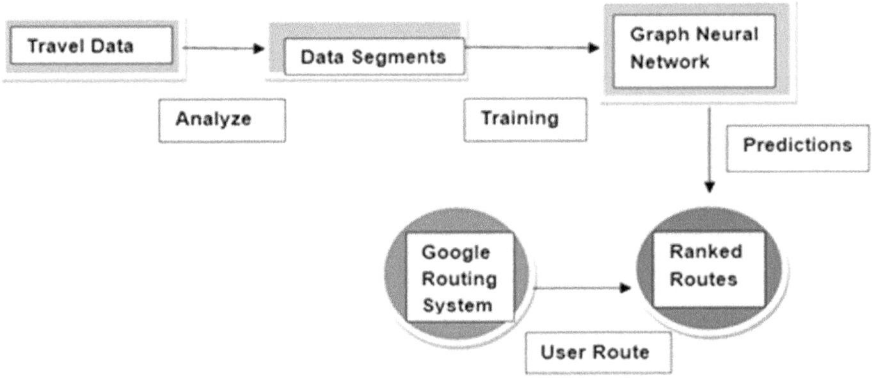

FIGURE 5.12 Google map usage of graph neural network.

5.10.2 RECURRENT NEURAL NETWORKS (RNNS) FOR DRIVING STYLE ANALYSIS AND VEHICLE DEMAND FORECASTING

The RNN is the type of deep neural network that is adapted for time series problems of variable length, permitting the use of past information to predict the future information.

5.10.2.1 DRIVING STYLE ANALYSIS

The RNN is used to analyze the driving style and GPS error like a vehicle in one place without any movements, internal sensor error and time gap based on the GPS data [9].

First, RNN model is trained using the preprocessed past GPS data (Accelerations and Speed). Based on the previous layer output, safety driving style evaluation is learned by the model. Then trained data is used to predict the future acceleration, speed factor and GPS error data for safety driving style. From the RNN driving style evaluation, we can instruct the drivers to improve their driving style for best performance and efficient personal driving [9].

The following parameters are considered for safe driving style:

1. Defensive driving;
2. Never drive tired;
3. Avoid distractions;

4. Stay alert;
5. Look ahead down the road;
6. Maintain optimum speed limit;
7. Wear Seatbelt;
8. Know your blind spots.

Different people will have different driving styles. So, the risk level varies for different driving styles. Identify and rate the drivers depending on their driving style.

5.10.3 *VEHICLE DEMAND FORECASTING*

Bidirectional gated RNNs are used for vehicle forecasting on demand. Each and every time, destination and departure demand on all stations are calculated. Both destination and departure values are summed and given to the activation function to calculate the predicted output. The model is learnt from current previous data (past hourly data).

5.11 APPLICATIONS OF RNN

Traffic speed, congestion flow and forecasting are predicted with the LSTM and gated RNNs.

5.11.1 *MULTILAYER PERCEPTRON IN TRANSPORTATION*

1. **Traffic Flow Prediction:** Triple layered multilayer perceptron is used to estimate the traffic volume and available parking place. These two factors play an important role in the smart city transportation system.
2. **Traffic Sign Recognition:** These systems play an important role in reducing the transportation risks. Traffic sign detection and detected sign recognition are the ideal steps for video traffic sequence.
3. **Multi-Layer Perceptron:** It is used for traffic sign recognition. Vehicle flow will not be the same in all places and time. So, training the multi-layer perceptron with numerous datasets is essential. If the received output is not the same as the predicted output, apply

the back propagation algorithm to learn from the past. Multi-layer perceptron model reduces error rate and gives accurate prediction even in the different traffic signs and lighting conditions.

5.11.2 VEHICLE CLASSIFICATION

The multilayer perceptron is used to classify the Vehicle. The first step in vehicle classification is detecting the vehicle through live camera capture, next extracting the feature from segmentation and the final step is classifying the vehicle based on pattern recognition (invariant features).

Viola-Jones algorithm is used for detecting the vehicle. The components of Viola-Jones object Detection algorithm are cascaded classifier, AdaBoost, Haar-Like feature and integral image. Haar-Like is a rectangular feature used as an input for cascaded classifier. In AdaBoost each cycle of boosting a feature among all other potential features is selected and, in the end, the final classification will be a linear combination of the initial weak classification. Integral image is a quick method for calculating the Haar-Like feature. For improving the processing speed and fast rejection of error windows, Cascaded classifier is used [10].

5.12 AUTOENCODERS AND LONG SHORT-TERM MEMORY (LSTM) IN TRANSPORTATION

5.12.1 DEMAND PREDICTION AND RARE EVENTS

The transportation should be increased to the place where the demand for the vehicle is high. Convolutional-LSTM is used for demand prediction based on spatial-temporal features taken from different input. Accident report is prepared after it happened. But if we can avoid the accidents in future by predicting accident features by autoencoder and traffic flow is modeled using long, short term memory (LSTM) [13].

5.13 SUMMARY

The neural network works like the human brain. The artificial neuron can communicate with other neurons and also learn from the past data to improve the accuracy in prediction and decision making. The transportation influences the country economy. The neural networks play the vital role to make the transportation as intelligent transportation system. In this chapter, neural networks computing capabilities and how neural networks make the intelligence in different transport applications is explored. The problems in the transportation are solved by using different neural networks.

Below neural network approaches can be used for various transportation applications like traffic forecasting, route planning, vehicle number plate detection, parking space identification, vehicle density, travel time, autonomous vehicle, vehicle location, vehicle speed, driver behavior, traffic management, etc.

1. **Convolutional Neural Networks (CNNs):** These are used in transportation applications like traffic characteristics and traffic incident risks prediction, vehicle identification, traffic signal timing, passenger demand-based transportation, AVs, traffic signal lights are changed based on the traffic, license plate recognition.
2. **Graph Neural Network (GNN):** These are applied to forecast the traffic and trajectory.
3. **Recurrent Neural Networks (RNNs):** These are used for driving style analysis and vehicle demand forecasting.
4. **Multilayer Perceptron:** It is used for traffic flow prediction, traffic sign recognition, vehicle classification.
5. **Demand Prediction and Rare Events:** These are done by autoencoders and LSTM.

KEYWORDS

- **convolutional neural network**
- **graph neural network**
- **Nesterov accelerated gradient**
- **recurrent neural network**
- **stochastic gradient descent**
- **support vector machine**

REFERENCES

1. Arya, K. H., Ravichandra-Mouli, V., Pranamesh, C., Yasaman, E., Saeed, A., & Anuj, S., (2020). Applications of deep learning in intelligent transportation systems. *Journal of Big Data Analytics in Transportation*. https://doi.org/10.1007/s42421-020-00020-1.
2. Deepika, S., Neeraj, K., & Prashant, S. R., (2020). Convolutional neural networks for 5G-enabled intelligent transportation system: A systematic review. *Computer Communications, 153*, 459–498.
3. Sorin, G., Bogdan, T., Tiberiu, C., & Gigel, M., (2019). A survey of deep learning techniques for autonomous driving. *Journal of Field Robotics*. Online ISSN: 1556-4967. https://doi.org/10.1002/rob.21918.
4. Sonata, I., Heryadi, Y., Lukas, L., & Wibowo, (2020). Autonomous car using CNN deep learning algorithm. *Journal of Physics: Conference Series: 2nd Annual Conference of Science and Technology* (Vol. 1869). https://doi.org/10.1088/1742-6596/ 1869/1/012071.
5. Krishnasamy, L., Ramasamy, T., Dhanaraj, R., & Chinnasamy, P., (2021). A geodesic deployment and radial shaped clustering (RSC) algorithm with statistical aggregation in sensor networks. *Turkish Journal of Electrical Engineering & Computer Sciences, 29*(3).
6. Weiwei, J., & Jiayun, L., (2021). Graph neural network for traffic forecasting: A survey. *Journal of LATEX Class Files, 14*(8).
7. https://deepmind.com/blog/article/traffic-prediction-with-advanced-graph-neural-networks (accessed on 1 January 2023).
8. Ramakrishnan, V., Chenniappan, P., Dhanaraj, R. K., Hsu, C. H., Xiao, Y., & Al-Turjman, F., (2021). Bootstrap aggregative mean shift clustering for big data anti-pattern detection analytics in 5G/6G communication networks. *Computers & Electrical Engineering, 95*, 107380.
9. Samuel, W., & Ulrich, G., (2020). Driving style analysis using recurrent neural networks with LSTM cells. *Journal of Advances in Information Technology, 11*(1).
10. Tarig, A., & Zaw, Z. H., (2016). Vehicle classification system using viola jones and multi-layer perceptron. *The International Arab Journal of Information Technology, 13*(6A).
11. Irina, V. P., Denis, A. P., Joel, J. P. C. R., Deepak, G., Ashish, K., Shankar, K., Changho, S., & Gyanendra, P. J., (2020). Automatic vehicle license plate recognition using optimal k-means with convolutional neural network for intelligent transportation systems. *IEEE Access*.
12. Gaurav, D., Supreeth, S., Neha, K. R., Amruthashree, R. V., & Eshitha, D., (2019). Traffic management using convolution neural network. *International Journal of Engineering and Advanced Technology (IJEAT)* (Vol. 8, No. 5S). ISSN: 2249–8958.
13. Matthew, V., & Medhat, M., (2020). Deep learning for intelligent transportation systems: A survey of emerging trends. *IEEE Transactions on Intelligent Transportation Systems, 21*(8).
14. Teresa, P., (2016). Neural networks in transportation research–recent applications. *Transport Problems, 11*(2).
15. Rajesh, K. D., Lalitha, K., Anitha, S., Khaitan, S., Gupta, P., & Goyal, M. K., (2021). Hybrid and dynamic clustering-based data aggregation and routing for wireless sensor networks [JB]. *Journal of Intelligent & Fuzzy Systems*, 1–15.

CHAPTER 6

Statistical AI Model in an Intelligent Transportation System

A. KODIESWARI[1], K. R. SABARMATHI[1], K. REMYA[2], N. R. GAYATHIRI[3], S. NITHYAPRIYA[3], and T. MALATHI[3]

[1]*Department of Information and Technology, Bannari Amman Institute of Technology, Erode, Tamil Nadu, India*

[2]*Department of Electronics and Instrumentation Engineering, Bannari Amman Institute of Technology, Erode, Tamil Nadu, India*

[3]*Department of Artificial Intelligence and Data Science, Bannari Amman Institute of Technology, Erode, Tamil Nadu, India*

ABSTRACT

The usage of vehicles has been increased over the past two decades due to increase in world population which leads traffic jamming, travel delays, road accidents and energy consumption. To solve this issue, intelligent transport system needs to be developed using powerful predictive model.

This chapter focuses on smart statistical AI models to address the issues of transport infrastructure development, traffic forecasting and driver behavior detection that avoids traffic congestion and accident. Neural network and artificial intelligence (AI) models provide an impact in the design of intelligent transportation model. The artificial neural networks (ANNs) are used to extract the patterns to develop transportation infrastructure systems (TIS). GNN ensures the estimated time of arrival by analyzing the traffic congestion to overcome traffic jams. Deep

Artificial Intelligence for Future Intelligent Transportation: Smarter and Greener Infrastructure Design. Rajesh Kumar Dhanaraj, Nilayam Kumar Kamila, Subhendu Kumar Pani, Balamurugan Balusamy, & Vani Rajasekar (Eds.)

Convolution Neural Network is used to detect the behavior of driver based on the characteristics like acceleration, gravity, throttle, and revs per minute (RPM) to reduce accidental risks. All these AI and neural network models bring safety to the civic, cost-effective, and reliable transportation system.

6.1 INTRODUCTION TO NEURAL NETWORKS

When building a neural network, the node layers are organized in a manner that is comparable to how neurons in the human brain interact with one another. Additionally, it is referred to as stimulated neural networks (SNN) or artificial neural networks (ANNs), and it is a critical component of machine learning (ML) applications. The node layers of ANNs comprises of three layers:

1. **Input Layer:** The very first layer of the neural network is the input layer. The data which needs to be processed will be given to the input layer. The input layer consists of the input nodes which will proceed the data to the subsequent nodes for further processing of the network.
2. **Hidden Layer:** There is a hidden layer in neural networks that sits between the algorithm's input and output, assigning weights to the inputs and using an activation function as an output to steer the network in the desired direction. To summarize, the hidden layers alter the inputs to the network in a nonlinear manner. The number of hidden layers has an effect on the performance of the neural network, and the weights assigned to those layers may have an effect on its performance. The hidden layers permit splitting down the functionality of the neural network into particular information transformations. Each function in the hidden layer is optimized to accomplish a certain result. Consider a neural network which needs to perform face recognition. The role of the hidden layer in this particular network is to split the process of face recognition to make it more effective. In order to identify the face using the network. It may split the process like one hidden layer may predict the eyes another may predict the nose on their own the network will independently recognize the objects, and they can operate together in a neural network to predict the whole object like face.

3. **Output Layer:** It may consist of the output nodes. The primary responsibility of this layer is to perform computations and transmitting information from the network to the outside world. The output from this layer only will be given to the user. The information from this layer will be in the form which could be understandable by the user.

6.2 WORKING OF NEURAL NETWORKS

Widely recognized as neural networks, ANNs and neural networks are two words that refer to the same idea. Using nodes, an artificial neural network may be divided into layers, such as the input layer, the hidden layer, and the output layer, which are then linked together to form the network. Each artificial neuron in the network that is linked to another neuron in the network is assigned a weight and a threshold. Whenever the output of a node reaches a certain threshold, the network deems the node to be active and transfers data to the next network layer. When the output node is less than the threshold value, then the data will not be permitted to move to the next layer. The prediction, classification, and clustering algorithms can be performed in the neural network by training the model with the available data set for a certain task. So, from the training data set, the neural network will learn, and this may help to increase the performance and accuracy of the network. Training data is a process used by neural networks to learn in order to increase their accuracy over time during the prediction. When these learning algorithms are tuned for accuracy, however, they become extremely powerful tools in the computer science and artificial intelligence (AI) domain, allowing anybody to easily categorize and organize data. When compared to manual categorization by human specialists, speech recognition and picture recognition operations can be done in minutes rather than hours. Google's search results are created using one of the most well-known neural networks in the world of networks (Figure 6.1).

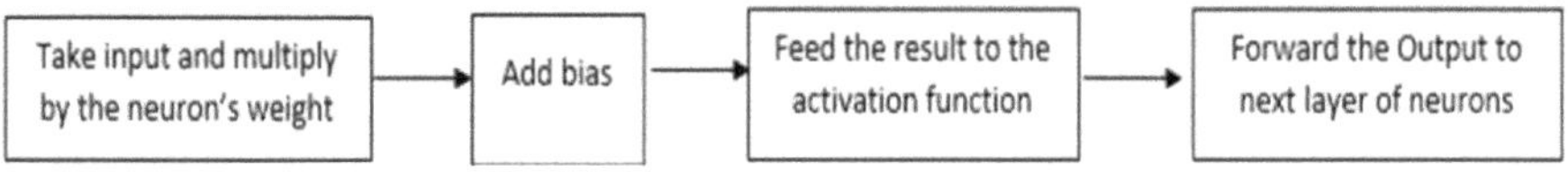

FIGURE 6.1 Working of neural networks.

Each and every node of a neural network will have its own input data, weights, a bias and an output. This can be estimated using the formula:

$$\sum_{i=1}^{m} wixi + bias = w_1x_1 + w_2x_2 + w_3x_3 + bias$$

$$Output = f(x) = \begin{cases} 1 \text{ if } \sum w1x1 + b \geq 0 \\ 0 \text{ if } \sum w1x1 + b < 0 \end{cases}$$

The process carried out in each node of the neural network is discussed below:

- Once the input layer receives the input data, the weights will be assigned to it.
- Input data will be multiplied with the assigned weights and then summed with a bias value.
- The obtained output of the node is supposed to pass through the activation function, and which will determine whether the output is capable of passing through or not.
- The initial threshold value will be fixed in the activation function. If the obtained output value exceeds the threshold, then the data of the node will fire, and if the obtained value is below the threshold value, then the data of the node will not fire.
- So, the output obtained from one node will be given as input for the node in the next layer.
- The neural network will be built by passing the data from one node to the next node.

Let us apply this in a real time example – Sheela wants to buy a dress. She is supposed to take a decision which may be either Yes (1) or No (0). To make the decision, Sheela needs to focus on a few influencing factors of the decision making. The factors are:

- Whether the dress fits her perfectly?
- Is the material of the dress good?
- Does the color suit her?

The network will check for these conditions before making a decision. Let us assume these as inputs. So, now we have three input parameters, and the input are:

- $X_1 = 1$, Because the dress fits her perfectly.
- $X_2 = 0$, Material of the dress is good.
- $X_3 = 1$, Color suits nicely.

Then, the weight for these inputs will be assigned according to the influence of that particular factor on the decision-making:

- W1-5, Since fitting is the major factor, it holds high weightage.
- W2-2, Material is the least factor considered by Sheela.
- W3-3, Color of the dress.

Where the weight is decided according to the priorities that Sheela prefers. Apply the values in the formula:

$$Y = (1 \times 5) + (0 \times 2) + (1 \times 3) - 3$$
$$Y = 8–3$$
$$Y = 6$$

So, the obtained output value is greater than the threshold value (which is 1). It indicates that the neuron can fire to the next layer now. If the value has less means, then the neuron is allowed to fire to the next layer.

In order to build a neural network, the following building blocks are required for handling of artificial neural network:

- Network topology;
- Adjustments of weights or learning;
- Activation functions.

6.2.1 NETWORK TOPOLOGY

The organization of a network's nodes and interconnecting links is referred to as its topology. ANN may be further can be categorized as follows based on their topology:

- feedforward neural network; and
- feedback neural network.

6.2.2 FEEDFORWARD NETWORK

It's a non-recurrent network with layers of processing units/nodes, all of which can be connected to the nodes of the earlier levels. Different weights are positioned on the connection. The signal can only go in one direction, from input to output, because there is no feedback loop. It may be classified into two categories (Figure 6.2).

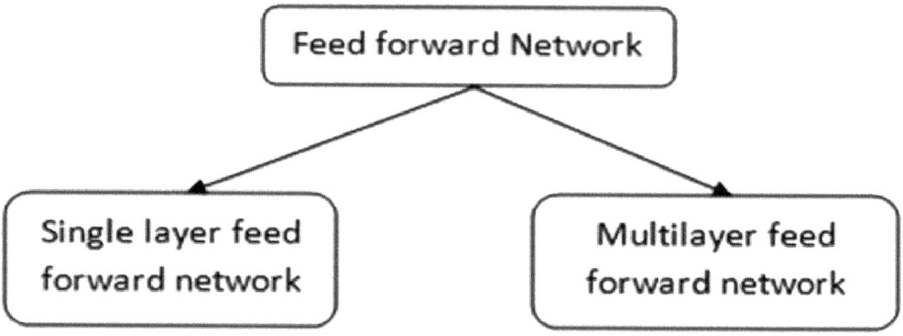

FIGURE 6.2 Types of feed forward network.

6.2.3 SINGLE LAYER FEEDFORWARD NETWORK

Single layer feed-forward network is the simple type of neural network which can establish a connection to the next layer neuron only. In this network, the cyclic connection between the nodes is not possible. Here, in this network data can be transmitted only in the forward direction. In a single layer feed forward network, there will be an input and output layers alone in the network. It will not have any hidden layers in between these layers (Figure 6.3).

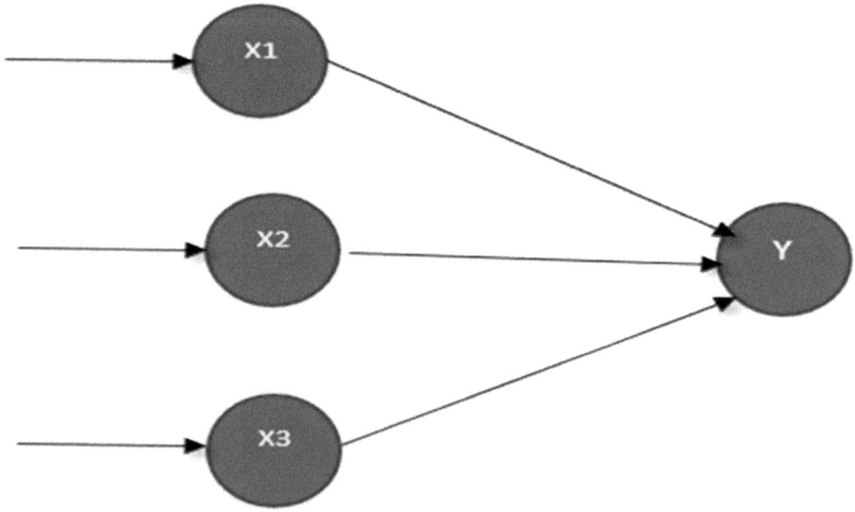

FIGURE 6.3 Architecture of single layer feed forward network.

6.2.4 MULTILAYER FEEDFORWARD NETWORK

Multilayer neural networks can be employed mainly on classification-based problems for non-linear types of data sets. In a multilayer neural network, the hidden layer will be included in between the input and the output layers of the network. The output of the input layer will be given as input for the hidden layer then the data will be processed to the output layer. There are no feedback connections, where the output of the network will be given as given again for the next iteration (Figure 6.4).

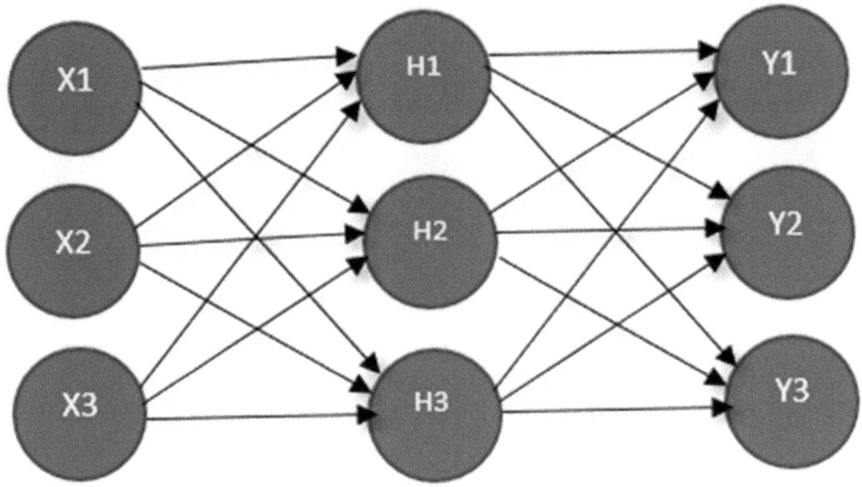

FIGURE 6.4 Architecture of multi-layer neural network.

6.2.5 FEEDBACK NEURAL NETWORK

In a feedback neural network, the data can be transmitted in both the forward as well as backward direction. The process will be carried out similar to the feed forward neural network like the input data will be multiplied with the assigned weight and summed with the bias. Here, the network will be sending the output as input for the next iteration and the process will be carried out until the network reaches the equilibrium point. So, we can usually consider the feedback neural network in a dynamic state. These networks can be used in addressable memories (Figure 6.5).

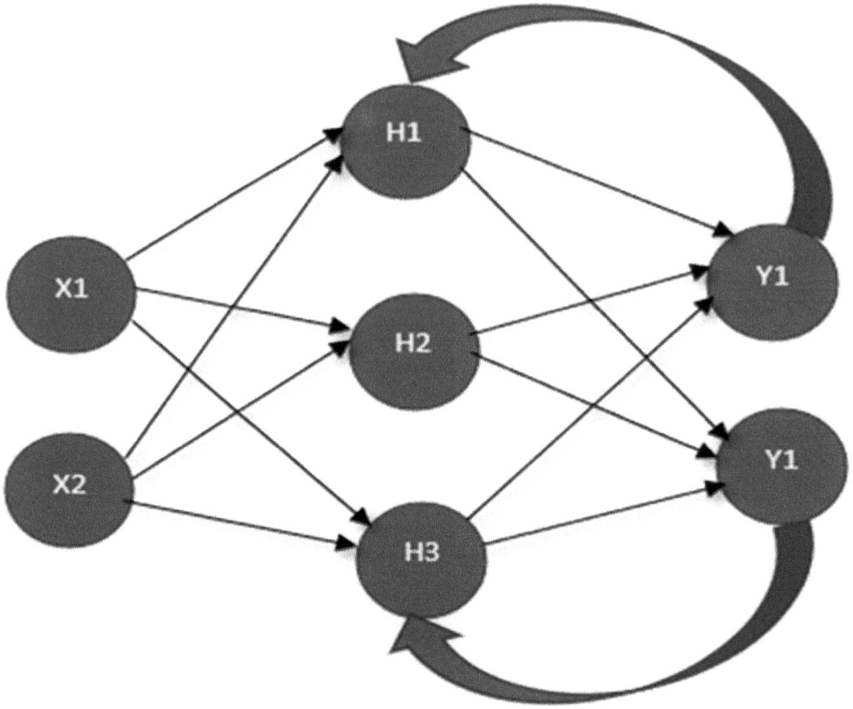

FIGURE 6.5 Architecture of back propagation neural networks.

A feedback neural network also called a recurrent neural network (RNN) can be a powerful and robust type of network used in sequential type of data or time series type of data. RNNs is an extended model of feed forward neural network. Generally, the RNN will recognize sequential characteristics of data and then use the patterns to predict the next likelihood using the recognized pattern. It is widely used in cases where prediction is difficult. The processed output data will be given as input for the previous state using a feedback loop. The feedback loop makes it persist and often described as memory.

6.2.6 ADJUSTMENT OF WEIGHTS

The weights in neural networks prove the strength of connection among two different neurons. It influences the system with appropriate input parameters. These weights are assigned to each node in the network. Bias are constant values that best fit the model with given input data. Once the model is trained with data, it should be further tested for accuracy. The error in the model is the difference between the actual and predicted values. Based on these errors, the system performs weight adjustments and propagates them backwards to the network. This process is called back propagation of error.

6.2.7 ACTIVATION FUNCTION

The activation function determines whether or not a neuron should be stimulated by calculating a weighted sum and then adding bias to it. Non-linearity is introduced into a neuron's output by using an activation function. Based on the output inaccuracy, we'd tweak the neurons' weights and biases in a neural network. This process is referred to as back-propagation. Activation functions enable back-propagation since the gradients and error are provided at the same time for updating the weights and biases. A neural network is nothing more than a linear regression model without an activation function. A non-linear activation function changes the input so that it can learn and do more difficult tasks. There are five important activation functions. They are:

- Linear activation function;
- Sigmoid activation function;
- Tanh activation function;
- ReLu activation function;
- SoftMax activation function.

6.2.7.1 LINEAR ACTIVATION FUNCTION

This function is represented by the equation $y = mx + c$, it is similar to the equation of a straight line. The linear activation function $f(x)$ can be represented as:

$$f(x) = x$$

The ultimate activation function of the last layer is nothing more than a linear function of the input to the first layer if all layers are linear in nature.

6.2.7.2 SIGMOID ACTIVATION FUNCTION

This function receives any real value as input and provides any value between 0 and 1 as outcome. The sigmoid activation function is shown below:

$$f(x) = \frac{1}{1+e^{-x}}$$

The closer the output value is to 1.0, the larger the value of the input (more positive), and the closer the output value is to 0.0, the bigger the value of the input (more negative). Models where the output is a probability prediction commonly employ this. This is due to the fact that the probability is somewhere between 0 and 1. This function is suitable in the range 3 to –3, after which the graph becomes smoothened to zero [15].

6.2.7.3 TANH ACTIVATION FUNCTION

This function is nearly identical to sigmoid/logistic activation in that it has a comparable output range of –1 to 1, and even has the same S-shape. Tanh Function is represented by:

$$f(x) = \frac{(e^{x} - e^{-x})}{(e^{x} - e^{-x})}$$

The output will be closer to 1.0 if the input is greater (more positive), and the output will be closer to –1.0 if the input is smaller (more negative). Because the tanh activation function's output is Zero-centered, we may easily translate the output values as highly negative, neutral, or very positive.

6.2.7.4 RELU ACTIVATION FUNCTION

ReLu is an abbreviation for Rectified Linear Unit. It is accomplished as a derivative function rather than a linear function. It is represented by:

$$f(x) = max\ (0, x)$$

It allows a feedback network model to convert the system into an efficient model. The ReLU function does not simultaneously stimulate all of the neurons. These neurons are turned inactive, when the output of the transfer function is less than 0. It is more efficient than sigmoid and tanh activation functions. Leaky ReLu is an improved version of the ReLu function. It provides the backpropagation for positive as well as negative values.

6.2.7.5 SOFTMAX ACTIVATION FUNCTION

The SoftMax activation function calculates the relative probabilities. The function is represented as:

$$softmax\ (z_i) = \frac{exp(z_i)}{\sum_j exp(z_i)}$$

The function is appropriate for multi-class classification. It returns the largest value in the list of values. The SoftMax function returns a probability of class membership for each class label and tries to approximate the predicted target for a given input as closely as possible.

6.3 MATHEMATICAL MODELS FOR TRANSPORTATION RESEARCH

Transportation research techniques were developed in the early 1960s. These methods were invented with a minimal number of samples or data. In this modern era of computing, large amounts of samples can be collected, stored, and retrieved. The main challenge would be to deal with the nonlinearity that can be seen in the data. Most of the models are linear statistical models.

Now the transportation research field is moving towards a drastic development in terms of efficiency. The data which is abundant in the current modern era can be used to identify patterns in them. It is also capable of dealing with noise and even uncertainties. One of the most important objectives is to collect large volumes of data in the transportation domain. Communication systems and sensor technology plays a vital role in data extraction from the real world. The efficacy of these technologies helps in improving the system to identify appropriate patterns in the data. Neural networks pave the way for finding various hotspots in transportation research.

Computational neural network replicates the functionalities of human intelligence. There are plenty of Computational Neural Network models available to solve the real-world issues in the transportation domain. Let us consider the feedforward neural network to express it with a clear mathematical model. Various technologies emerged in recent times helps in improving the computation efficiency of the systems. Research on computational, physics, cognitive science, statistics contributes to the development of neural models in depicting human intelligence. These intelligent systems are capable of processing information that the human brain processes.

A neural network is made up of nodes that are interconnected and work similar to neurons in the human brain. The deep learning algorithms can be applied to recognize the pattern in given data, classify the data, cluster them and so on. Neural networks are a computational model which holds a network architecture made up of artificial neurons. They have applications in various areas like search engines, recommendation systems, smart personal assistants and much more.

Neural networks have its hindrance in domains like e-commerce, Logistics, Healthcare, Finance, Security. It is widely used for personalizing the user's experience in shopping, predicting the best route for travel, examining patient's health, monitoring the patients' health status periodically, detecting fraudulent actions, forecasting the stock market growth, and protecting from system attacks. Hence, Neural networks are used in almost every place wherever intelligence is adopted.

The main objective of neural networks is to transform the input into significant output. The network is made of artificial neurons which take multiple inputs and process them to generate a worthwhile outcome from the system. Each neuron has an influence on the other neurons. It is capable of observing the data with complex patterns. Neural networks have many layers. Individually each layer proposes to perform a specific function. Hence, the model of a neuron in a neural network is termed to be a perceptron. The perceptron takes several inputs and processes it to produce a single binary output. A neuron typically consists of input values, weights, transfer function, activation function, and output value. The weights are real numbers that depict the importance of the input feature. Figure 6.6 shows the structure of a neuron.

In the given figure, input values (X_i) indicate the input feature values that are fed to the model to train the model for better accuracy. These features are to be chosen appropriately, such that they are closely related to the task to be accomplished. The most important feature is provided with more weights (W_i) to increase their contribution to the learning process. Certain features create a great impact in deciding the outcome of the task. For instance, the road condition creates a great impact in the travel timing between given spots for any traveler. A threshold is set for the entire process. The output is determined by the weighted sum of the dot product of all the input features and their corresponding weights. The target output is 0, whenever the weighted sum is less than the chosen threshold value. Also, The target output is 1, whenever the weighted sum is greater than the given threshold value.

$$Output, Y = \begin{cases} 0, \sum X_i \; W_i < threshold \\ 1, \sum X_i \; W_i > threshold \end{cases}$$

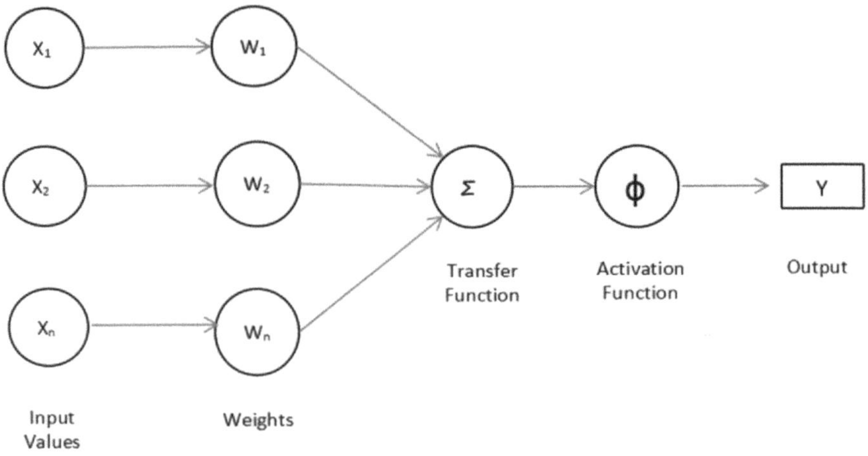

FIGURE 6.6 Structure of a neuron depicting the human brain.

6.4 ALGORITHMIC NEURAL NETWORK MODEL FOR TRANSPORTATION RESEARCH

In a neural network, node layers are arranged in a manner similar to how neurons in the human brain communicate with one another. Neural Networks is also known as Artificial Neural Network (ANN) and Stimulated Neural Network (SNN) and it is said to be subset of Machine Learning and Deep Learning algorithm. There are three types of node layers in ANNs: the input layer, a hidden layer, and an output layer.

The weights and thresholds of each artificial neuron linked to another artificial neuron are the same. An individual node is engaged when its output exceeds a predetermined threshold value, and data is sent to the next tier of the network. When the output node is less than the threshold value, no data is sent. Classification, clustering, function approximation, time-series analysis, and forecasting problems like these are all solved with neural network.

- Traffic forecasting and control;
- Evaluation of traffic parameters;
- Maintenance of transport infrastructure;
- Transport policy;
- Driver behavior and autonomous vehicles (AVs).

Best neural network techniques to meet the transportation research challenges are discussed as in subsections.

6.4.1 ARTIFICIAL NEURAL NETWORKS (ANNS) FOR TRANSPORTATION INFRASTRUCTURE SYSTEMS (TIS)

With their exceptional capability to infer referring from complex or erroneous data, ANNs are used to extract complex patterns from data and recognize drifts that are too complex to be detected by people or some other intelligence-based approaches. The concept of ANN was presented by Zhang & Zhang (1999), and it was created to imitate the functioning and structure of neural systems in living beings. ANNs is considered to be a powerful tool for various applications like medicine, transportation, optimization, quantum physics, text classification, language generation, speech recognition and so on.

ANNs are made up of a huge number of neurons, which are connected in parallel and work together to solve various issues.

Input and target data are used to train the ANN. ANNs also illustrate the whole interconnections of the models along with the numeric weights of the nodes that may be changed depending on experience, allowing neural nets to adapt to inputs and learn. Thus, ANN is considered to be a powerful and flexible tool for seizing non-linear and complex physical process with a high degree of accuracy.

6.4.1.1 ADVANTAGES OF ARTIFICIAL NEURAL NETWORK

- Works with large set of data amount of data set;
- Detects complex nonlinear interactions between dependent and independent variables implicitly;
- Indicate each and every potential connection amongst the predictor variables;
- Also solve complex problems in TIS (transportation infrastructure systems).

6.4.1.2 DESIGN OF TRANSPORTATION INFRASTRUCTURE SYSTEMS (TIS)

Transportation infrastructures are defined as the interconnection of physical and perceptible assets necessary to sustenance and develop a country. It is consequently critical to effectively manage the Transportation Infrastructure in order to deliver continuous, long-term, and cost-effective amenities to the population. The design of TIS is a critical component of effective administration.

Many competing variables are currently affecting the Transportation sector, necessitating rigorous system considerations like high design and construction capital costs, legislative changes, and the influences to increased requirement and utilization of Transportation Infrastructure. Planning land usage, building construction, and land subdivision are all parts of transportation infrastructure planning and development that are supervised by the governments at all levels. The goal is to produce high-quality and long-term development outcomes in both urban and rural areas, and also accounting crucial problems like environmental preservation as part of a larger Transportation Infrastructure Planning and Development analysis. For effective development of TIS, the system engineering methodologies are highly required.

Transport infrastructure engineering requirements consists of four core elements, and they are (Figure 6.7):

- Infrastructure asset rehabilitation and renewal;
- Structural performance and operation;
- Sustainable and tangible materials;
- Emerging robust technologies and innovative tools.

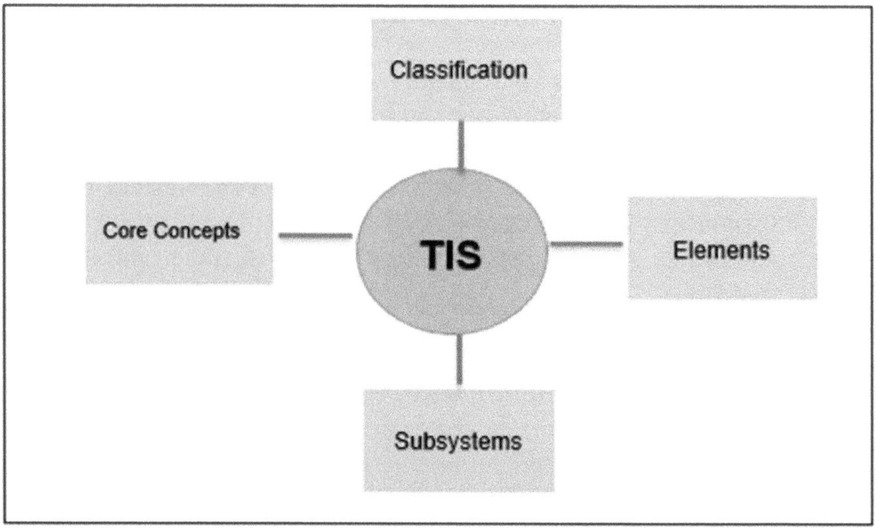

FIGURE 6.7 TIS parameters.

The general system engineering methods involves:

$$TX_n = \Sigma WQ_n = OX_{nq}$$

where; TX_n is the transportation system parameters input; W is the TIMS weights; Q_n is the function; and OX_{nq} is the output of TIMS functions.

6.4.1.3 TRANSPORTATION INFRASTRUCTURE MANAGEMENT SYSTEM (TIMS)

The straightforward approach to improve the performance of transport infrastructure systems (TIS) is transportation infrastructure management system (TIMS). TIMS is a highly specialized method that works with a wide range of transportation characteristics. TIMS consists of two subsystems, and they are pavement management system (PMS) and bridge management system (BMS). The objective of TIMS is to collection of various sizable data related to transportation infrastructure. To improve the performance of TIMS, a conditional enhancement method is used. Infrastructure damage detection and monitoring is another challenging task for TIMS since the visual evaluation strategy is used traditionally, which consumes more time and is difficult in structure diagnosis of the state. This affects the performance of TIMS. To overcome, an ANN approach is used which involves structural parameter reduction for damage detection and diagnosis of constituent material state for monitoring process.

6.4.1.4 ARCHITECTURAL ANN DESIGN OF TRANSPORT INFRASTRUCTURE SYSTEMS

When it comes to data mining prediction and classification, ANNs are a data-driven tool (Figure 6.8).

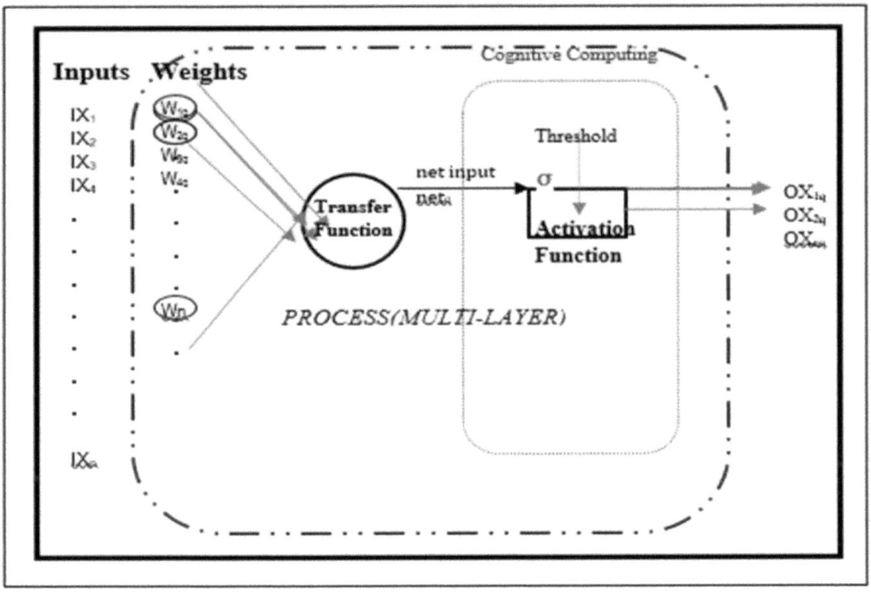

FIGURE 6.8 Artificial neural network function or transportation infrastructure system.

Furthermore, ANNs provide unique tools and applications like Control devices that, whenever implemented appropriately, may provide substantial advantages in sectors like Transportation Infrastructure. The suggested model is a dependable prototype with additional benefits, especially in terms of Process Control and System Performance predictions. Because process control and system performance are regarded essential components of TIS, TIS, and TIMS must be included in any artificial Neural Network collaboration. TIS ANN is a multilayer general system engineering solution. Method includes input, processing, and output phases. This model's overall performance must be consistently improved and maintained in order to aid in the overall development of priorities. Furthermore, all three main system phases must be properly studied and integrated holistically when evaluating system performance enhancement.

6.4.2 TRAFFIC FORECASTING USING GRAPH NEURAL NETWORK (GNN)

Traffic prediction and estimated times of arrival (ETA) are the important factors to overcome traffic jams and running to late hours. Traffic forecasting helps friends or family members to intimate the late arrival and also the service providers to pass the change in pick-up or drop timing information. Google Map computes the ETA by analyzing the traffic data from road segments using ML algorithms. The ML algorithm analysis the future traffic by comparing the live traffic data with the historic road traffic patterns. But it is complex to compute the exact traffic in roads due to inevitable variations in rush hours. Also, other added challenges for predictive models are road quality, speed limits, accidents, and closures.

6.4.2.1 WORKING PRINCIPLE OF GRAPH NEURAL NETWORKS (GNNS) IN GOOGLE MAPS

The working principle of the existing ML model for traffic forecasting resides on super segments, but estimating the travel times with super segments are found to be difficult. Multiple neighboring segments of road that share significant traffic volume are referred to as "super segments." Super segment covers to a group of road segments, each of which has its own length and speed characteristics. To achieve a high percentage of accuracy, GNNs, a ML model was utilized in recent studies. GNNs, which incorporate relational learning biases for spatiotemporal reasoning, to simulate the connection structure of real-world road networks. GNN will design a traffic model as how cars flow through a network of roads (Figures 6.9–6.11).

The super segment covers a set of road segments in which each segment has particular length and speed features. Initially the RNN was used to handle variable length sequences of super segments and found difficulty in forming the structure from the road network. The RNN is deployed by the concept of "proximity" and allows complex connectivity to handle traffic ahead, behind us, adjacent, and intersection roads. GNN is recommended for traffic modeling, and it is an extended learning inclination of CNN (convolution neural network) (Figure 6.12).

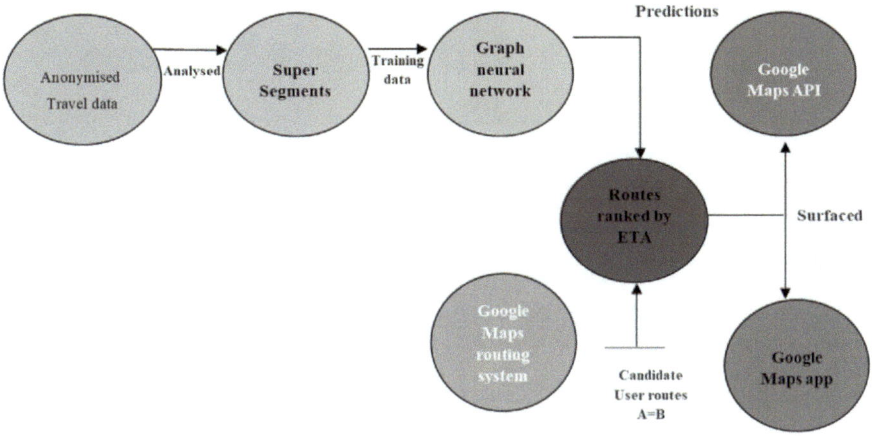

FIGURE 6.9 The model architecture for determining optimal routes and their travel time.

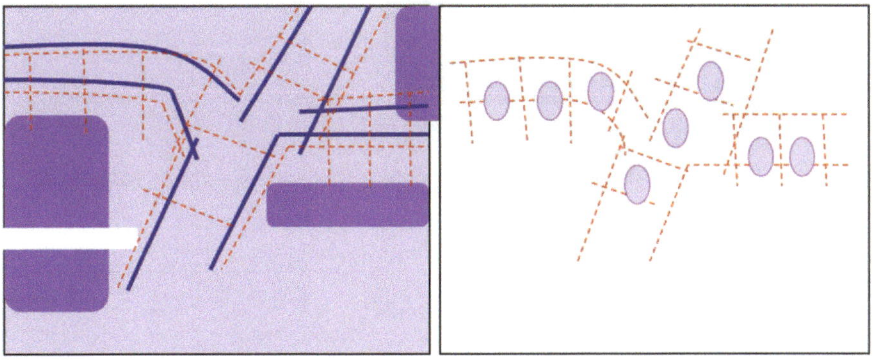

FIGURE 6.10 Working principle of advanced graph neural networks in google maps.

In a GNN, the local road network is treated as a graph. It also includes the node for each route segment and consecutive edges of the same road or associated through a crossroads. In our model, the states of nodes and result

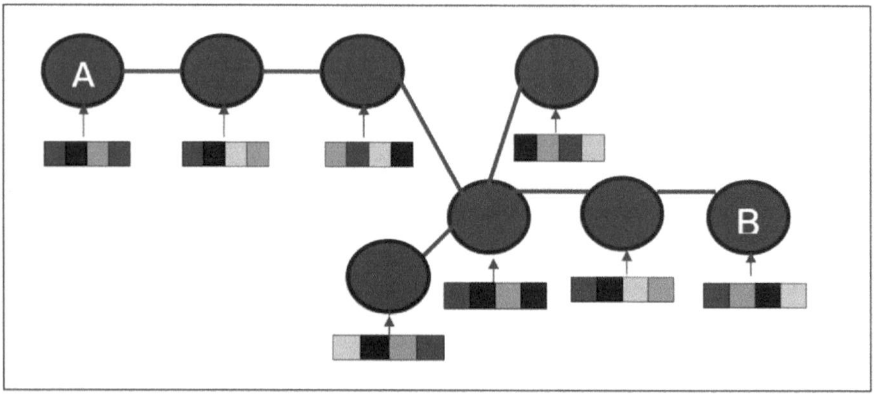

FIGURE 6.11 Functions of recurrent neural network.

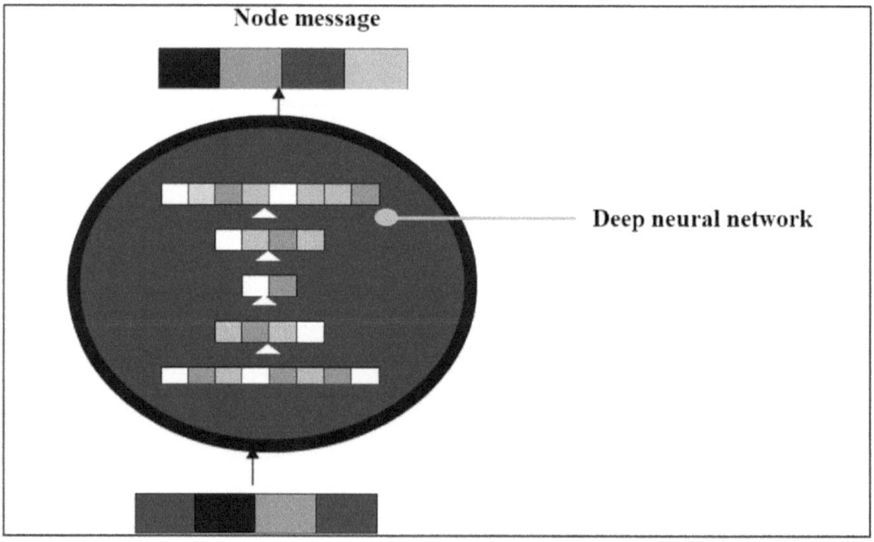

FIGURE 6.12 Working of deep neural networks.

of edges are determined by neural network when a message passing algorithm is executed. So, it is understood that super segments are considered as road networks sub graphs, which were sampled at random in proportion to traffic density. These sampled sub graphs are trained using single models and deployed at a system. In GNN model the adjacent nodes send messages to one another and by appointing local bias the nodes will find confidential adjacent nodes with one message passing step. This system enables the GNN to develop the connectivity structure for the effective road networks. The adjacent roads not connected with the main road also predicted accurately which makes GNN more powerful experimentally. Nonetheless single route or long route, GNN performs the same process according to varying length and complexity.

6.4.2.2 BUILD GENERALIZE MODEL

Since there are so many variations in the graph topologies utilized during training, when the model is put into production, it becomes extremely sensitive to changes in the training curriculum. A single batch can include anything from a two-node tiny network to a huge graph with more than 100 nodes [11]. The constraints of GNN were solved by utilizing a revolutionary reinforcement learning (RL) technique known as supervised learning. When a ML model learns anything new, it forgets essential features it has already learnt. As a result, the learning rate of the model is reduced over time by the researchers. After a predetermined time of training, the parameters are normalized using an exponentially declining learning rate schedule.

To overcome the error in learning rate during training runs, MetaGradients RL algorithm is used to obtain optimal learning rate dynamically during training runs. The Meta Gradients readjust the learning rates during training runs, enable to reduce the learning rate and further improve the quality at a greater rate automatically. Hence this approach is most widely used in production as it induces more stable result.

Generic models with great potential to reduce travel estimate errors are built using a linear combination of multiple loss functions. We used a regularizing factor on the model weights, L2 and L1 losses on the overall traversal times, and individual Huber and negative-log likely (NLL) losses for each node in the graph to produce a multi-loss target [12]. We avoided over-fitting on the training dataset by including these losses into our model. Training advancements were more quickly transmitted to held-out test sets and end-to-end trials even though our judgments of training quality were unchanged.

6.5 DEEP CONVOLUTIONAL NEURAL NETWORKS (CNNS) FOR DRIVER BEHAVIOR DETECTION

Convolutional neural networks (CNNs) is an ordinary approach that uses deep learning techniques and neural network architectures with more than one hidden layer. Deep Neural Network is a model with multiple hidden layers. CNNs are also known as deep neural networks. Transportation is a primary mode of economic impact for any developing countries and recently there is drastic increase in the number of vehicles on road. Due to increase in vehicles the road safety is a severe problem around the world. Hence to reduce accidental risk it is responsible to detect the driver behavior using ITS (intelligent transportation systems). The traditional computer vision techniques suffered with spoofing and privacy violation issues. By utilizing deep learning algorithms, it is possible to examine driver behavior using a variety of characteristics, including as acceleration, gravity, throttle, and revolutions per minute (RPM). When using the criteria, the approach can determine if a driver is being safe, aggressive or distracted while driving. It may also determine whether the driver is intoxicated while driving. The value of deep neural networks on pictures may be utilized by using CNNs on images to build models from driving signals using recurrence plot (RP) approaches (Figure 6.13).

The deep CNN techniques for driver behavior detection involve:

- Data acquisition;
- Implementing time window and recurrence plot technique;
- Convolution neural networks.

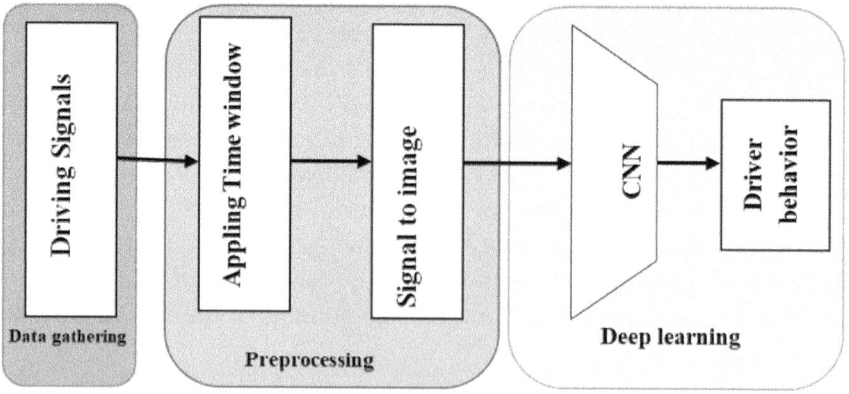

FIGURE 6.13 Process of driver behavior detection.

Time window is applied on 9 collected signals to differentiate the driver behavior. The window data are then converted into image using RP technique. The converted image enables to convert temporal dependencies into spatial dependency. The images that are generated will be given as input to CNNs to classify different driving styles such as normal, aggressive, distracted, drowsy, and drunk driving. The efficient and low-cost computational model will be designed based on number of convolution layers, number of filters and filter size (Figure 6.14).

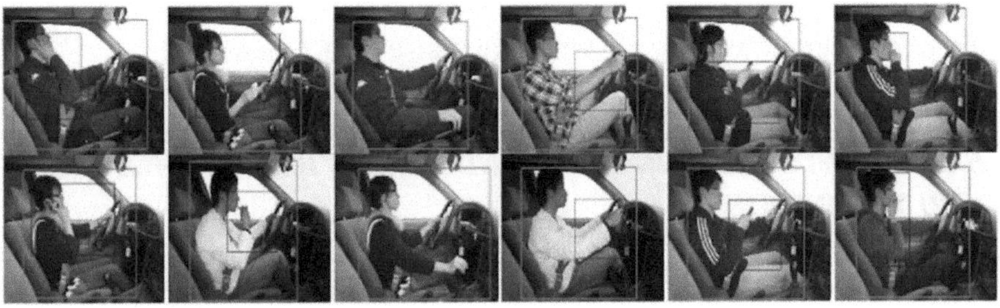

FIGURE 6.14 Different driving style from SEU dataset.

6.5.1 DATA ACQUISITION

We require the vehicle's linear acceleration regardless of the vehicle's orientation to examine its movement. The linear acceleration is a measurement of the pressure applied to a vehicle along all three axes (x, y, and z), except the force of gravity. To compute linear acceleration, we need both acceleration and gravity inputs in all three directions (x, y, and z). We consider acceleration and gravity to be two properties that the deep neural network is tasked with evaluating and learning representations for.

Revolutions per minute is a unit of measurement for the number of times an engine rotates around a fixed axis in one minute. ECU parameters must be accessed via an OBDII adapter in order to acquire an accurate RPM value. According to how firmly the accelerator pedal is pressed, the throttle position sensor decides how far the throttle valve will open up. The OBDII adaptor retrieves this value from the ECU. RPM changes in proportion to throttle in most, but not all, conditions, such as uphill or downhill routes. This is especially true in manual (non-automated) cars. As a result, while making predictions about driving behavior, these features are taken into account.

A driver was simulated who engaged in aggressive behaviors such as speeding, tailgating, excessive acceleration, and slowing down when taking off or braking. A fake activity was used to draw the driver's attention away from the road for 1 to 2 seconds, simulating distracted driving.

To mimic sleepy driving, participants closed their eyes for a random period between 1 and 2 seconds, seldom applied the brake, had sluggish steering, drove slower than the speed limit, and had minimal change in acceleration or gear. A sleepy driver, in reality, will ultimately wake up and either correct the vehicle's position or cause an accident.

6.5.2 IMPLEMENTING TIME WINDOW AND RECURRENCE PLOT (RP) TECHNIQUE

We record the behavior of drivers in each time slot, which is classified as normal, aggressive, inattentive, drowsy, or inebriated. The 5 s time frame is used for detecting drunk driving, a 20 s time window for detecting aggressive driving and a 1.8 s time window is used for identifying distracted driving. Data is gathered in the time domain and is dependent on time. The recurrence plot (RP) is a time series visualization tool to visualize recurrent tools.

The mathematical model of recurrent plot is:

$$R_{i,j}^{X} = \theta\,(\epsilon - \|X_i - X_j\|)$$

where; $\|.\|$ is the norm; x is the time series; θ is the Heaviside function; and ϵ is the recurrence threshold.

6.5.3 CONVOLUTIONAL NEURAL NETWORKS (CNNS) ARCHITECTURE

An engine's revolutions per minute (RPM) is a unit of measurement. ECU parameters must be retrieved using an OBDII adapter. With a throttle position sensor, you can precisely control how much the throttle valve opens when you depress the accelerator pedal all the way down. The OBDII adaptor collects data from the ECU and sends it to the computer. However, this is not always the case. For example, the number of RPM fluctuates in relation to throttle in uphill or downhill routes, and especially in manual (non-automated) cars. Because of this, these qualities are taken into account when making predictions about a driver.

CNN architecture is thought to be the most accurate and computationally efficient model for detecting driver behavior. The adoption of a novel CNN architecture eliminates the need of a large number of learnt parameters in a very deep fully connected network. Instead, at the conclusion of CNN, we use a very tiny fully linked network with global max pooling. These new CNN trends demonstrate outstanding efficiency with minimal parameters in numerous applications, and they are also less prone to overfitting. The efficiency and quicker convergence features of the leaky ReLU activation function are used (Figure 6.15).

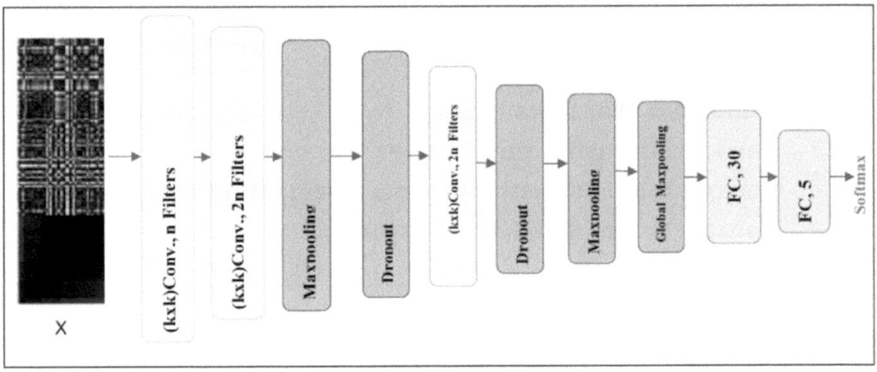

FIGURE 6.15 Convolution neural networks architecture.

All input images for CNN's three channels are 150 × 50 pixels in size. In the initial layers of the CNN, the convolutional and max pooling layers function as feature extractors, at the same time the SoftMax classifier classifies the input pictures into one among five categories in the final layer. Dropout is a normalization technique that reduces overfitting by randomly disregarding certain neurons.

6.6 CONCLUSION

This chapter discussed on AI models to develop Intelligent Transport System to overcome the confrontation in transport system. The statistical

AI models for Transport Infrastructure Management, Traffic forecasting and detection of driver behavior was explained with an application [13]. Based on the performance analysis, AI is considered to be an effective model in the design and development of optimal networks for transportation system. Additionally, it helps in the prediction of travel demands, weather-based travel plan, and current traffic state and also diversion management during road accidents. The chapter provides the future scope on CO_2 emission, safe travel, pollution control alert and efficient fuel utilization.

KEYWORDS

- **artificial neural networks**
- **machine learning**
- **stimulated neural networks**
- **transport infrastructure systems**
- **transportation infrastructure management system**
- **transportation infrastructure systems**

REFERENCES

1. Iyer, L. S., (2021). AI enabled applications towards intelligent transportation. In: *Transportation Engineering* (Vol. 5, p. 100083). Elsevier BV. https://doi.org/10.1016/j.treng.2021.100083.
2. Damaj, I., Al Khatib, S. K., Naous, T., Lawand, W., Abdelrazzak, Z. Z., & Mouftah, H. T., (2021). Intelligent transportation systems: A survey on modern hardware devices for the era of machine learning. In: *Journal of King Saud University – Computer and Information Sciences*. Elsevier BV. https://doi.org/10.1016/j.jksuci.2021.07.020.
3. Wang, J., Chen, R., & He, Z., (2019). Traffic speed prediction for urban transportation network: A path based deep learning approach. *Transportation Research Part C: Emerging Technologies*, 372–385.
4. Abduljabbar, R., Dia, H., Liyanage, S., & Bagloee, S. A., (2019). Applications of artificial intelligence in transport: An overview. In: *Sustainability* (Vol. 11, No.1, p. 189). MDPI AG. https://doi.org/10.3390/su11010189.
5. Silva, N., Shah, V., Soares, J., & Rodrigues, H., (2018). Road anomalies detection system evaluation. In: *Sensors* (Vol. 18, No. 7, p. 1984). MDPI AG. https://doi.org/10.3390/s18071984.

6. Kujala, R., Aledavood, T., & Saramäki, J., (2016). Estimation and monitoring of city-to-city travel times using call detail records. *EPJ. Data Sci.*, 1–16.
7. Karlaftis, M. G., & Vlahogianni, E. I., (2011). Statistical methods versus neural networks in transportation research: Differences, similarities and some insights. *Transp. Res. Part C Emerg. Technol., 19*, 387–399.
8. Boyuan, Y., Yongli, Z., & Jie, Z., (2020). *Demonstration of Joint Optimization Between Cloud AI and On-board AI in Optical Transport Networks*. ISSN: 2166-8892, IEEE.
9. Indrabayu, Rizki, Y. B., Intan, S. A., & Ais, P. A., (2017). *Vehicle Detection and Tracking Using Gaussian Mixture Model and Kalman Filter*. doi: 10.1109/Cybernetics Com.2016.7892577, IEEE.
10. Ken, A., Piers, C., & Alessandra, O., (2005). *Hybrid Decision Support Based on Knowledge Discovery and AI Techniques for the Management of Maintenance Services in the Public Transport Sector*. doi: 10.1109/IDAACS.2005.283071, IEEE.
11. Donghoun, L., Sunghoon, K., Sehyun, T., & Hwasoo, Y., (2019). Real-time feed-forward neural network-based forward collision warning system under cloud communication environment. *IEEE Transactions on Intelligent Transportation Systems, 20*(12), 2019.
12. Kan, G., Yongli, H., Zhen, Q., Hao, L., Ke, Z., Yanfeng, S., Junbin, G., & Baocai, Y., (2021). Optimized graph convolution recurrent neural network for traffic prediction. *IEEE Transactions on Intelligent Transportation Systems, 22*(2).
13. Can, L., Lei, B., Wei, L., Lina, Y., & Travis, W. S., (2020). Graph neural network for robust public transit demand prediction. *IEEE Transactions on Intelligent Transportation Systems*, 1–13.
14. Shaofei, W., (2021). Spatiotemporal dynamic forecasting and analysis of regional traffic flow in urban road networks using deep learning convolutional neural network. *IEEE Transactions on Intelligent Transportation Systems, 2021.*
15. Kan, G., Yongli, H., Zhen, Q., Yanfeng, S., Junbin, G., & Baocai, Y., (2020). Dynamic graph convolution network for traffic forecasting based on latent network of Laplace matrix estimation. *IEEE Transactions on Intelligent Transportation Systems, 2020.*
16. Xuelong, L., Zhiyuan, Z., & Wang, Q., (2021). ABSSNet: Attention-based spatial segmentation network for traffic scene understanding. *IEEE Transactions on Cybernetics, 11*, 1–11.
17. Ling Zhang & Bo Zhang (1999). A geometrical representation of McCulloch-Pitts neural model and its applications. In *IEEE Transactions on Neural Networks* (Vol. 10, Issue 4, pp. 925–929). Institute of Electrical and Electronics Engineers (IEEE). https://doi.org/10.1109/72.774263 (accessed on 4 March 2023).

CHAPTER 7

Reinforcement Learning in Smart Transportation

S. MALATHY[1], C. N. VANITHA[2], and RAJESH KUMAR DHANARAJ[3]

[1]*Assistant Professor, Computer Science and Engineering, Kongu Engineering College, Erode, Tamil Nadu, India*

[2]*Assistant Professor (Sr. G), Computer Science and Engineering, Kongu Engineering College, Erode, Tamil Nadu, India*

[3]*Associate Professor, Computer Science and Engineering, School of Computing Science and Engineering, Galgotias University, Uttar Pradesh, India*

ABSTRACT

Artificial intelligence (AI) and machine learning (ML) are currently trending self-learning technologies in the numerous industrial applications. While the majority of existing technology and knowledge are focused on two methodologies as follows: (i) supervised learning; and (ii) unsupervised learning. But the true AI paradigm as presented by nature is the capacity to learn while engaging with the situation.

7.1 SUPERVISED LEARNING

Traditional supervised learning is improbable, because not at all real-world creatures would have been confronted with a set of positive and negative examples. Living beings, on the other hand, learn via interaction and

Artificial Intelligence for Future Intelligent Transportation: Smarter and Greener Infrastructure Design. Rajesh Kumar Dhanaraj, Nilayam Kumar Kamila, Subhendu Kumar Pani, Balamurugan Balusamy, & Vani Rajasekar (Eds.)

experimentation. They are learning not only a model to differentiate between distinct groups, but also the best policy to achieve the intended result.

Supervised learning is appropriate in the process of learning generalizations to new-fangled situations, but it will not be appropriate in the process of learning via interacting with the environment. The database contains all conceivable interactions, each of which is labeled with the appropriate action. The agent can generalize by inferring a course of action from an unknown environment condition, but it will not find a way of correcting itself, if it is incorrect.

7.2 UNSUPERVISED LEARNING

This technique analyzes and clusters unlabeled datasets using machine learning (ML) techniques. Without the need for human intervention, the algorithms in this technique will uncover data groupings or hidden patterns.

It is the best solution for cross-selling techniques, exploratory data analysis, image identification and consumer segmentation because of its capacity to detect similarities and differences in information.

7.3 REINFORCEMENT LEARNING (RL)

The process of learning tells how to relate conditions to actions which will maximize a reward value numerically is known as RL [1]. The system is not taught which actions to take; instead, it must attempt them all to see which ones provide the best results.

RL is a type of semi-supervised learning methodology that falls under the category of policy learning. There are three important factors to be considered in RL as illustrated in Figure 7.1:

- reward;
- policy; and
- actions.

7.3.1 POLICY

A policy describes how an agent acts at a specific point in time. A policy, in general, is a mapping between environmental conditions with the actions and the activities, the agent makes in the environment. In the simplest circumstances, the policy can either be a simple mathematical function or sequence of actions listed in a table, but it can also be a complex computation. Policy is represented as the foundation of the knowledge base (KB) for the agent.

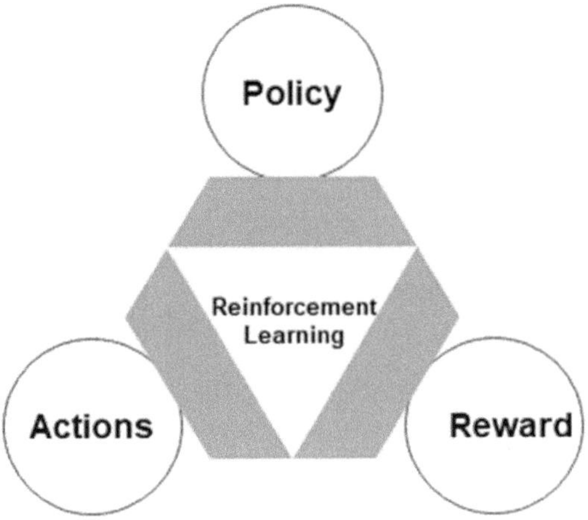

FIGURE 7.1 Factors of reinforcement learning.

7.3.2 REWARD

The goal or an objective of a RL problem is defined by a reward. The agent's actions result in a reward at each time step. The agent's ultimate goal is to maximize the reward he or she obtains. As a result, the reward distinguishes between the agent's positive and negative action outcomes. Both pleasure and pain experiences are rewarded.

The primary approach to influence the policy is through the reward; if a policy-selected action results in a lower reward value, then the policy can be altered to select a different deed in the same circumstances.

7.3.3 ACTION OR VALUE FUNCTION

A value function describes what is desirable in the long run, whereas the reward signal shows positive activities in an instant sense: each action results in an immediate reward.

The state value is defined as the whole amount of rewards that an agent can obtain from the beginning state. Values represent the enduring attractiveness of a collection of states, taking into consideration the most probable states for the future as well as the sequence of rewards derived from them. In spite of a state provides a modest immediate reward, it might still be valuable since it is frequently followed by states that provide bigger benefits.

7.3.4 IMPORTANCE OF TAKING DECISION ON THE VALUES AND REWARDS

- Rewards are the primary values which will be obtained immediately after applying an action in RL;
- Values are secondary projections which follows the obtained rewards.

Each reward is associated with the values. The sole point of calculating values is to obtain additional rewards. Nonetheless, values are considered when making and assessing decisions. Value judgments are ultimately used to guide action decisions. The agent will select the activities that bring the uppermost value states, not the uppermost reward, because those states will initiate the actions that will obtain the most reward in the elongated term.

7.3.5 ENVIRONMENT

The environment provides rewards directly, but values are constantly approximated based on tracing the sequences of observations that the agent makes during each and every encounter. As a result, the most significant component of RL algorithms will be the method(s) for effectively estimating values.

7.3.6 MODEL OF THE ENVIRONMENT

The model of the environment is another crucial component of some RL systems. This is something that replicates environmental behavior and provides assumptions regarding the environment's behavior. This kind of model will assist an agent in predicting the next reward if a particular action is made, allowing the agent to base current action selection on future environment reactions.

7.4 KEY PRINCIPLES IN REINFORCEMENT LEARNING (RL)

Figure 7.2 illustrates the key principles of RL.

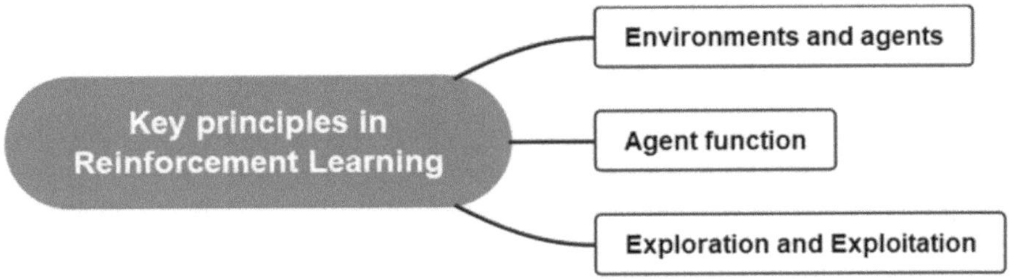

FIGURE 7.2 Key principles of reinforcement learning.

7.4.1 ENVIRONMENTS AND AGENTS

- An agent is defined as an object that senses its surroundings with the help of sensors and reacts to them through actuators.
- Sensors include eyes, ears, and other organs, whereas actuators include legs, wings, and other body parts.
- A software agent accepts sensory inputs and responds to them by creating files, exhibiting information on the screen and sending packets through the network.
- At any one time, an agent's choice of action is determined by what he perceives or has perceived up to that point, but not by what it hasn't perceived.

7.4.2 AGENT FUNCTION

- The agent's behavior is denoted by a function which contains the mapping of an action to the situation.
- By mapping the agent's reaction to every sequence, the agent function that translates each sequence to an action is defined.

7.4.3 EXPLOITATION VS. EXPLORATION

In order to increase overall performance, a RL agent would progressively learn the optimal (or near-best) strategy based on brainstorming, considering the reactions of random interactions with the environment. This is possible for an agent by means of two ways as follows:

1. **Exploration:** Learning better strategies.
2. **Exploitation:** Exploiting the situation which means greedily using the best strategy available.

Exploration is expensive in terms of resources [6], opportunity, and time, a key concern in RL is how to balance exploration into uncharted terrain with the use of current proven tactics. In particular, the agent must balance between what he has learnt thus far by applying greedy exploitation and choosing activities that currently provide the uppermost reward, and uninterruptedly exploring the situation to obtain more information in order to possibly reach a higher long-term value.

➢ **Example 1:**
Assume the area to be cleaned is splitted into 9 grids of a square with coordinates G and H, where the values of G = 1, 2, 3 and H = 1, 2, 3 as shown in Figure 7.3. The vacuum agent can tell the grid it is present and also whether the grid is dirty or not. It has the option of moving up, down, left or right, sucking up the dirt, or idle, i.e., without doing any action.

(1,1)	(1,2)	(1,3)
(2,1)	(2,2)	(2,3)
(3,1)	(3,2)	(3,3)

FIGURE 7.3 Grid view of a square for vacuum cleaning problem.

➢ **Agent Function:** The following is a representation of an agent function:
- If the present grid of a square is unclean, the agent has to initiate the action of sucking;
- Otherwise, it has to move on to the next grid of a square.

➢ **Reward:** It's critical to specify when an agent receives a reward and whether it's positive or negative. Giving a constructive reward every time the agent scrubs all of the squares is a naive technique. However, because the agent explores at random, there is a slim probability of getting the prize by cleaning all of the squares.

A better technique for guiding the agent in order to achieve the desired goal is to provide a little constructive reward all the time if the vacuum agent scrubs a square and a slight bad reward if the vacuum agent tries to clean a grid of a square that has already been cleaned. When all of the squares have been cleaned, there will be a significant positive reward.

- **Value Function:** The vacuum cleaner problem would be solved by utilizing the value function. Each number represents the most recent approximation of our chances of finishing the cleanup in that state. This estimate is referred to as the state value, and the entire table is referred to as the learned or informed value function.

  State A has a greater value is determined as "better" than state B if the present estimate of our chances of winning from the state A is higher than the state B. Because if all of the space has been cleaned, all scenarios in which all of the squares have been cleaned have a probability of 1.

  To choose the next move, the agent looks up the current values of the states from each possible motion as follows.

  - **Function 1:** To denote the one of the four directions;
  - **Functions 2 and 3:** Agent scrubs the dust or not.

In most of the attempts, the agent would select the move which results in attaining state possessing the highest value, which means that one with the maximum projected likelihood of identifying an unclean square. Sometimes the agent may choose at random from among the other possible movements. Such moves are referred to as exploratory moves, which allow us to encounter states that will not be encountered otherwise. When we add a reward based on how quickly the space is cleaned to the value function, then the agent would be able to choose a better space traversal approach.

7.5 ALGORITHMS FOR REINFORCEMENT LEARNING (RL): AN OVERVIEW

The goal is to present an intuitive overview of known RL methods without delving too deeply into the models or the math behind them. RL is a more sophisticated and difficult way to implement, but it essentially deals with:

- Learning through collaboration and feedback;
- Learning to complete a task through trial and error; or
- Obtaining rewards by applying actions in an environment.

In essence, an agent (or several agents) is created that can sense and understand the circumstances in which it is situated, as well as act and interact with it.

7.5.1 BASIC TERMINOLOGIES

The basic terminologies involved in RL [3] is shown in Figure 7.4:

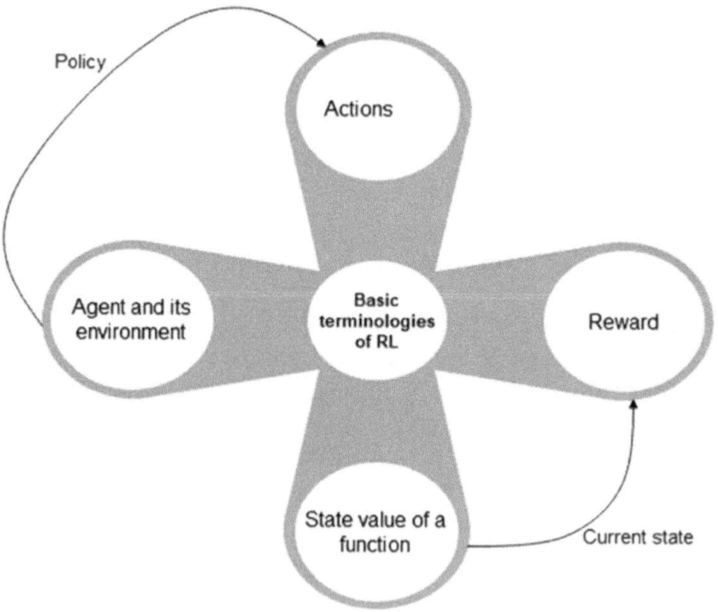

FIGURE 7.4 Basic terminologies of reinforcement learning.

- The learner and decision-maker are the agent.
- The agent's environment is where he or she observes, listens, and then decide what type of actions to be taken next.

- A collection of actions that the agent can carry out.
- The agent's current status in the environment.
- The environment delivers a reward to an agent for each, and every action performed by it. A scalar value is frequently used.
- Policy is the agent's supervisory function (decision-control strategy), which will map the situations to actions.
- The value of a state is the enduring reward obtained by the beginning state and performing a specific policy.

7.5.2 CATEGORIES OF REINFORCEMENT LEARNING (RL) ALGORITHMS

- The challenge of inducing a function from training samples is referred to as the function approximator. Examples for common approximators are as follows:
  - o Neural networks;
  - o Nearest-neighbor;
  - o Decision trees algorithms.
- A probabilistic model:
  - o A precise number of states are present in this approach.
  - o The present state and action chosen will be useful for estimating the probability distribution for the future states.
- Dynamic programming (DP):
  - o It is a group of problem-solving techniques for consecutive decision problems with a composite cost structure.
- Monte Carlo methods:
  - o It is a type of value function learning method.
  - o It is used for evaluating the state value by performing many attempts from the beginning state and then averaging all the rewards obtained so far on those attempts.
- Temporal difference algorithms:
  - o It is a type of learning algorithm that compares temporally sequential predictions. Perhaps the most basic concept in all of RL.
- The agent's perception of the world is represented by a model that maps state-action pairings to probability estimations over states.

- **Example 2:**

  We have an agent and a prize, but there are a lot of roadblocks along the way. The agent's job is to find the most efficient route to the reward. The difficulty is better explained in the following problem.

  The robot's objective is to obtain the diamond prize while avoiding the fire obstacles. The robot learns by attempting all feasible courses and then selecting the one that provides the best reward with the fewest obstacles. Each correct step will reward the robot, while each incorrect step will deduct the robot's pay-out. When you approach the final prize, the diamond, the entire reward will be computed.

  The most important aspects of RL are presented in Figure 7.5.

1. **Input:**
   - The input should be a starting state for the model to work from.
2. **Training:**
   - Depending on the input, the model will suggest a state;
   - The user will select whether to give a reward or punishment based on the output and the learning from the previous experience;
   - The model is constantly evolving;
   - The optimal solution is determined by the highest possible payment.
3. **Output:**
   - There are numerous possible outputs, just as there are numerous solutions to a given problem.

The goal of RL is to make judgments in a sequential manner. In simple terms, the output is determined by the current input's state, and the next input is determined by the previous input's output. The initial input or the input delivered at the start is decided in supervised learning. Because decisions in RL are reliant on one another, we assign labels to sequences of dependent decisions, whereas in supervised learning, each decision is given its own label.

- **Benefits of Reinforcement Learning:**
  - Enhances performance;
  - Change can be sustained for a long time;
  - Increases behavior;
  - Disobey the basic standard of performance.

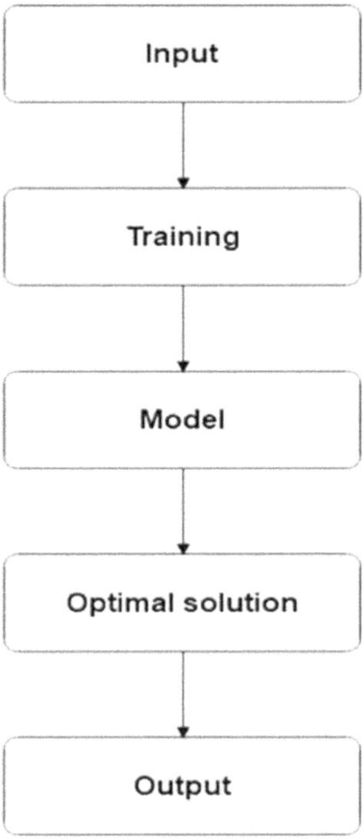

FIGURE 7.5 Aspects of reinforcement learning.

- **Drawbacks of Reinforcement Learning:**
  - Applying too much of reinforcement might lead to an overabundance of states, lowering the quality of the output;
  - It only gives enough to meet the bare minimum of requirements.
- **Applications of Reinforcement Learning:** The various applications of RL [5] is given below:
  - Robotics;
  - Industrial automation;
  - Data processing;
  - Constructing training systems.

7.5.3 SMART TRANSPORTATION IN SMART CITIES

Smart transportation in smart city management system will be designed to handle the cities by incorporating emergency response along with mobility while minimizing traffic congestion on the streets. The integration of sensors, high-speed networks and modern communication technologies with automation are responsible for the progress of smart transportation. The world is entering into new stage of smart transportation called the intelligent or smart transportation systems (STS) and the internet of things (IoT) [7].

7.6 INTELLIGENT TRAFFIC MANAGEMENT

"Intelligent and smart transportation networks use a variety of technologies for not only managing the transportation systems but also to monitor and evaluate the same [2], which in turn will improve the level of safety as well as efficiency" as shown in Figure 7.6.

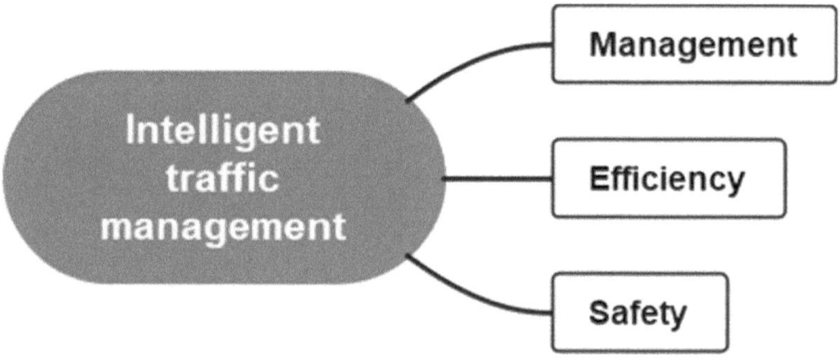

FIGURE 7.6 Intelligent traffic management and its parameters.

These systems are based on the three parameters stated below:

- Management;
- Efficiency;
- Safety.

To put it another way, smart transportation makes movement about a city in following dimensions:

- Cost-effective;
- Safer;
- More convenient.

7.6.1 WHAT NEW TECHNOLOGY ARE ENABLING THESE NEW POSSIBILITIES?

The spread of 5G communication technology with the IoT devices are the significant factors of STS. IoT devices are made of with inexpensive controllers and sensors on a physical machine which is maintained remotely. 5G technology provides the low-latency and the high-speed connectivity needed for smart and intelligent transportation network control and administration.

Smart transportation is already being implemented in a number of places, with the lessons learned from their triumphs and failures being applied to new systems. A few examples are listed below:

- Major hubs such as New York City have incorporated the innovation of smart transportation as part of their ongoing transformation into a smart city.
- The rural state of Wyoming, on the other hand, is a prominent testbed for linked automobiles not only increases the efficiency of the supply chain but also minimize the necessity for long distance traveling truckers who must balance the fully loaded schedules with the manpower for the rest.

7.7 THE MOST IMPORTANT ADVANTAGES OF TRANSPORTATION TECHNOLOGY

7.7.1 INFORMATION FROM INTELLIGENT TRAFFIC MANAGEMENT SYSTEMS

There are various advantages realized in smart city when it is enriched with smart transportation as presented in Figure 7.7.

7.7.1.1 SAFER

Autonomous transportation systems have the potential to eliminate the "human factor" in accidents by combining the technologies such as IoT, 5G and ML. Computers will not be easily tired, distracted or emotional.

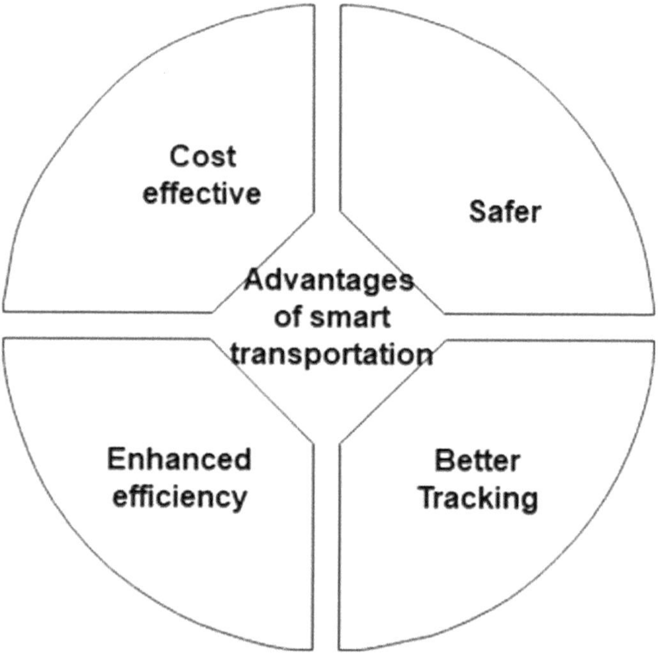

FIGURE 7.7 Advantages of smart transportation.

7.7.1.2 BETTER TRACKING ABILITY

Data gathering is a critical component of responsible infrastructure management. It not only offers complete data for every component of the intelligent transportation system, but it also aids managers in improved tracking maintenance requirements, identifying critical sources of problems and monitoring operations that must be handled.

7.7.1.3 ENHANCED EFFICIENCY

Better management translates to more efficient utilization. High quality data is more useful for not only recognizing areas but also for enhancing the efficiency. Perhaps a minor change in rail schedules would result in advanced fill rates, and if the bus stops are distributed, the bus routes will serve the neighborhood in a better way.

7.7.1.4 COST-EFFECTIVE

Since an intelligent transportation system is making an efficient use of available resources, results in reduced energy usage and also the costs, preventative maintenance, and minimal resources allocated to vehicular accidents. Riders can save money when public transportation is affordable, and it is efficient to strive with an action of owning the private vehicles.

7.7.2 TRAFFIC MANAGEMENT COMPONENT

They can receive quick prominence and notifications for accidental-prone and crowdy areas affecting public safety, emergency response systems, traffic on city streets [10], and allowing them to interact more effectually with emergency responders and other agencies.

Aside from the previously mentioned benefits of improved administration, safety, and efficiency, there are a number of other advantages [8] that the local governments, general public, and around the world which is listed below:

- Security;
- Resilience of the supply chain;
- Environmental points to consider.

7.7.2.1 SECURITY

The susceptibility of smart cities to cyber-attacks is a key concern for skeptics. After all, as the globe becomes more linked, cyber-attacks on key structure such as Internet-connected banking systems and power grids have become practically regular among criminals and even nation states [12].

Long before computers, electricity grids, and banks are other essential infrastructure, such as transportation, were susceptible to physical attacks. When network infrastructure and vehicles are integrated, the feature of autonomous is improved which increases the physical risks such as car theft, terrorists utilizing vehicles as weapons, and an intention of holding public transportation as hostages can all be eliminated with the use of cryptography [9].

Cyber-attacks, on the other hand, are far easier to fight against the physical threats mentioned above. Cyber-attacks can be mitigated through the techniques presented in Figure 7.8.

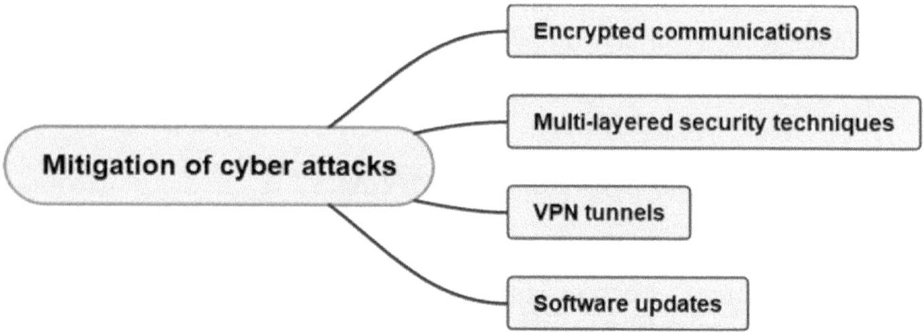

FIGURE 7.8 Techniques involved in the mitigation of cyber-attacks.

- Encrypted communications;
- Multi-layered security techniques;
- Virtual private network (VPN) tunnels;
- Software updates.

This means that smart transportation when incorporated in smart cities can make current public transit safer by minimizing the risk of physical attacks as well as cyber-attacks [13].

7.7.2.2 SUPPLY CHAIN RESILIENCY

A world-wide catastrophe like the COVID epidemic has demonstrated how sensitive the world's supply lines once they are interrupted. When the heavy and light vehicle drivers become unwell, the public health is at high risk once they are in transition from one place to another. In these situations, autonomous goods transportation can be used to save the lives of the people.

Projects like Wyoming's linked car project could pave the way for an autonomous supply chain. It is designed with intelligent city-to-city transportation cum logistics systems to transport crucial products like food and emergency supplies without putting human drivers at risk. Automation, artificial intelligence (AI), and robotics can be used to enhance the human workforce and supply chain with the improvement in safety and security.

7.7.2.3 ENVIRONMENTAL POINTS TO CONSIDER

Transportation's history is inextricably linked to the environment. Transportation has a negative impact on the planet's resources and atmosphere, from coal consuming steam cars to today's gas-based combustion engines.

An extent of scientific progress is being made each and every day to identify alternate energy resources needed for power transportation. Smart transportation also helps communities to more responsibly use their existing resources.

Although public or group-based transportation is healthier for the society than using the private vehicles, which will not be extensively used in the United States and in a few other countries due to its inconvenient nature in some areas. Smart transportation solutions, on the other hand, can increase the efficiency of modern public transportation, making it more appealing to a wider range of people. As urban transportation technology advances, successful approaches may be reproduced and expanded to regions that desire to reap the benefits of a smart transportation-based city.

7.7.3 DRAWBACKS OF SMART CITY TRANSPORTATION SYSTEMS

With all of its advantages, some drawbacks may emerge as a result of smart and intelligent based city transportation systems are adopted. These issues primarily revolve around two factors:

- Power usage; and
- Data management.

7.7.3.1 SENSORS AND POWER USAGE

Smart cities are designed with sensors and all these sensors necessitate power. This will necessitate the use of batteries for sensors linked to moving objects. While the fixed or immovable sensors is potential enough to run on solar power, they will almost always need to be connected to the electrical system of the smart city. It is practically impossible to power up the large number of sensors (Say trillion sensors) needed to convert the whole world into smart cities.

Even with sensors connected to the electrical grid, the quantity of raw material required (e.g.: copper) is far greater than what the global population is used to producing.

7.7.3.2 DATA MANAGEMENT

Aside from electricity, there is a lot of discussion about the utilization of internet by a mass population which results in a huge volume of personal data online these days.

Data is acting as the building block or backbone of smart cities. Without it, they would not be able to function. While the majority of the data required is unstructured and unauthenticated in comparison to online data. It will necessitate a behavioral as well as the mental shift among those people. Cars should acquire the positioning data whereas the sensors placed across a city has to passively capture the signals emitted by a smartphone during the day.

If the smart cities integrated with smart transportation needs to prosper in the future, responsible rules and procedures for managing data should be developed.

7.8 WORKING PROCEDURE OF SMART TRANSPORTATION

7.8.1 VEHICLE-TO-VEHICLE COMMUNICATIONS IN TRAFFIC CONTROL

Public infrastructure and the automotive sector are the two primary categories that smart transportation falls into. When the sensors connected in a network are linked into both the infrastructure and on the vehicles to achieve the goals of remote monitoring and vehicle control and personal safety, these two sectors become "smart."

Consider a congested city junction. Pedestrians are attempting to cross the street. The traffic flow is controlled by the lamps. Vehicle drivers are frantically attempting to reach their destination. Traditional transportation systems use timers or pressure plates beneath the road or placing the buttons on the curb for the pedestrians to turn on the lighting.

It is the responsibility of both drivers and pedestrians to pay attention to (and obey) traffic lights. However, if any one of the nodes in the network fails, both safety and efficiency suffer. A driver who is inattentive runs a red light. When a pedestrian forgets to press the crossing button, they miss their turn and must wait longer.

All of this, however, changes at a smart intersection. A vehicle can identify pedestrians using a mix of Bluetooth and LiDAR and automatically begin brakes to prevent an accident. Pressure plates and timers cannot pick up individual signals transmitted by automobiles, so streetlights can identify not only the exact number of cars are waiting in a road but also the direction more accurately and effectively. Cars and streetlights can even interact to the point that the car starts moving when the light becomes green, and slows down and stops when the light turns red.

All of this is made possible by the use of technology in transportation, such as the IoT and 5G connectivity speeds enabling real-time activities and remote sensing.

7.9 URBAN TRANSPORTATION TECHNOLOGY EXAMPLES

7.9.1 A BUSTLING METROPOLIS

Regardless of the benefits and drawbacks of smart cities, the technique is now available and in use. Smart sensors and adaptive controllers are being used in passenger-information systems [11], public transportation dispatch systems and in train networks all around the world and in major parts of the United States. Here are some instances of smart city transportation.

7.9.1.1 SMART TRAFFIC MANAGEMENT IN MIAMI DADE

Southern Pennsylvania Transportation Authority (SEPTA) is a public transportation agency in the state of Pennsylvania (SEPTA) train control with a positive attitude (PTC).

7.9.1.2 SMART DISPATCH SYSTEM (SUBURBAN MOBILITY AUTHORITY FOR RAPID TRANSIT)

7.9.1.2.1 Advanced Traffic Management System (ATMS) of Miami Dade

With a population of nearly 2.5 million people, Miami Dade County is Florida's most populous county. The County of Miami Dade is responsible for managing traffic flow throughout the urban region that makes up the city of Miami and its neighboring areas, including the operation of nearly 2,700 signalized intersections. In fact, every year, the number of signalized intersections and mid-block crossings grows by dozens.

The Miami Dade ATMS, which integrates Digi 4G LTE cellular routers as a chunk of the communications structure in county-wide traffic cabinets, is intended to minimize traffic congestion and delays while also increasing mobility.

7.9.1.3 PTC SEPTA

SEPTA (Southeastern Pennsylvania Transportation Authority) is in charge of Philadelphia's light rail, subway, and bus systems. With over 1 million daily riders, these services must be dependable and safe every time a vehicle leaves the station. SEPTA created a positive train control (PTC) system to signal trains, prevent derailments and crashes, and track speed and signal infractions.

With the Digi WR44-RR mobile access router, SEPTA achieve this. This device, when installed aboard a train, enables for remote communications with roadside sensors through a radio link. The device provides signals with train movement data while also receiving information about road closures and other circumstances that can force a plan adjustment. This prevents trains from speeding into a potentially dangerous situation.

7.9.1.4 SMART DISPATCH SYSTEM

It is a computer-assisted dispatching system. In smart cities, public transportation is a crucial area for connected technology innovation. Over

300 buses are managed and dispatched by SMART (Suburban Mobility Authority for Rapid Transit Authority) in Detroit.

Buses are a key component of how people travel around, therefore they must be on time, safe, and free of breakdowns. The city employed three radio towers in an analog radio network, which is dispersed throughout the region to facilitate bus dispatch and location tracking.

They used the Digi WR44 R mobile cellular router when it was time to upgrade. The transition from analog to digital enabled much better tracking and administration. A new innovation allowed SMART to track not just the position of each car, but also its speed and maintenance data.

For society, the smart transportation market holds a lot of promise. People will realize efficiency, cost benefits and safety which comes with contemporary public transit as intelligent transportation systems (ITS) for smart cities become more widely used around the world. It's fascinating to consider how civilization will engage with their cities in the future, given the new technology that are becoming available.

7.10 TRANSPORTATION REINFORCEMENT LEARNING (RL)

RL is used to tackle problems requiring consecutive judgments. It works through trial-and-error learning a strategy over time. This structure appears simple, yet it is really sophisticated, and unexpected behavior can develop. In my new book, RL in detail, you may learn more about the specifics of RL.

Transportation, in particular, faces a wide range of these difficulties. Take, for example, driverless vehicles, which is now a separate area. Moving anything or someone from one location to another is essentially a large sequential decision-making problem.

ML, on the other hand, creates models from data. RL is no exception, and it frequently requires additional input because it must explore in order to find optimal solutions, which implies that in all but one situation, behavior is sub-optimal and sometimes catastrophic. There are a variety of technical techniques for promoting safe RL, but most studies start with simulations. Even after proving in the real world, returning to the simulation with new data and learning is beneficial.

But what are the most prevalent transportation issues being addressed?

1. **Self-Driving Cars:** Autonomous driving presents a number of challenges that are well-suited to RL. Sensing the environment, leveraging information from other vehicles, decision-making, strategic planning, and job execution are all common responsibilities. A lot of work has gone into direct vehicle control, which is commonly used in limited circumstances such as on the highway or for constrained activities like autonomous lane shifting.

   Other than direct control, some of the most exciting studies look into other topics. The inverse RL to learn the reward function of an average human driver, rather than attempting to specify a reasonable reward function up front, which is difficult.
2. **Routing and Electric Vehicles:** Electric vehicles (EVs) face extra issues because they require frequent recharging, and the number of recharging sites is restricted. Researchers have explored into using graphs and RL to redistribute EVs for business fleets, such as those used in logistics. Short-term car rental companies have similar challenges.
3. **Control of Traffic Signals:** All modes of automotive transportation are constrained by signaled intersections. The notion is that the control optimizes vehicle flow, however most existing implementations rely on static logic and extremely simplified traffic distribution assumptions. It's been proven time and time again that RL outperforms traditional traffic signaling techniques.

   RL agent to optimize overall traffic flow and would be far more effective than the current human techniques.

   RL presents the option of automating low to no-risk merging of vehicles coming from slip roads in the future, when required speeds can be signaled to individual automobiles or drivers. Even today, where slip roads have some minimal kind of regulation, such as traffic lights, this is conceivable.
4. **Various Modes of Transportation:** RL can be used to improve a variety of modes of transportation, not just those on the road. Train scheduling on both the surface and underground is an evident sequential decision-making task, where rescheduling, re-routing, or canceling trains can help relieve track congestion while minimizing passenger delays. Train shunting and automatic train operations are both scheduling issues, with the extra complexity of determining train speeds and locations. One user who uses RL for shunting activities is the Dutch train company NS.

As a result, it should come as no surprise that another application field is air traffic control. The main reason for this is that manually operating and controlling such a large number of flying objects is getting increasingly challenging. With the growing number of low-altitude drones and vertical takeoff and landing vehicles, a more complex and automated solution, such as RL, is becoming more important.

However, you might be surprised to learn that another prominent RL topic is ship automation. The intersection of two key RL applications, scheduling, and control, is represented by shipping. Because shipping is a commercial activity that exists due to logistical requirements, optimizing the timing and routing of goods to arrive to a destination makes sense. Autonomous ships have lately been studied to increase safety, lower expenses, reduce environmental effect, and combat piracy. Because of their ability to manage complicated and dynamic systems, RL agents may be trusted with the control and navigation of these ships.

7.11 FINAL THOUGHTS ON TRANSPORTATION EQUALITY

There are lot of issues in directing transportation which are listed as follows:

- Minimize delay;
- Maximize flow;
- Safety issues;
- Lack of data updation in real-time.

RL paves a way to an entirely different approach to how transportation infrastructure is prioritized:

- Transportational access to high-quality institutions such as schools, hospitals, libraries, and even markets are linked to educational and health outcomes.
- Differing racial and socioeconomic groups have different levels of access to metropolitan transportation networks.
- Governmental agencies can intervene to enhance facilities in underserved areas or to improve transit connections.

It could be as simple as modifying a bus route in terms of implementation with RL to increase social equality by optimizing bus routes. RL has the potential to improve the lives of those who would otherwise lack the ability to shape their own destiny.

RL is a cutting-edge ML technique that optimizes sequential decision-making. It has the potential to change the way we travel, move things, and, hopefully, improve the lives of people who are less fortunate. If you're interested in learning more about RL, I highly recommend my book "Reinforcement Learning: Industrial Applications of Intelligent Agents."

Agents are accomplished with reward and punishment system in RL. The agent is rewarded for good decisions and penalized for bad ones. The agent seeks to minimize the number of incorrect moves while increasing the number of correct ones.

7.11.1 SELF-DRIVING CAR APPLICATIONS

Deep RL has been proposed in a number of articles for autonomous driving. There are several factors to consider in self-driving automobiles, including speed limitations in various locations, drivable zones, and avoiding crashes, to name a few.

Motion planning, trajectory optimization, scenario-based learning policies, controller optimization and dynamic pathing for highways are few autonomous driving activities where the technique of RL could be used.

Learning automated parking policies, for example, can help with parking. Q-Learning can be used to change lanes, and overtaking can be done by implementing an overtaking procedure or policy while evading collisions and maintaining a constant speed afterwards.

The AWS DeepRacer is an autonomous racing automobile that was created to put RL to the test on a real-world track. It controls the throttle and direction using a RL model and cameras to visualize the runway.

Wayve.ai has successfully used RL to teach an automobile to drive in just one day. To solve the lane following problem, they developed a deep RL algorithm. They used a deep convolutional network with four convolutional layers and three fully linked layers.

7.11.2 REINFORCEMENT LEARNING (RL) FOR INDUSTRY AUTOMATION

Learning-based robots are utilized to execute many jobs in industry reinforcement. Apart from being more efficient than humans, these robots are also capable of performing activities that would be dangerous for humans.

Deepmind's usage of AI agents to cool Google Data Centers is a wonderful example. This resulted in a 40% reduction in energy consumption. The AI system currently controls the centers completely without the need for human involvement. Data center experts are evidently still in charge of supervision. The following is how the system works:

- Taking five-minute snapshots of data from data centers and feeding them to deep neural networks;
- It then forecasts the impact of various combinations on future energy usage;
- Identifying measures that will result in minimal energy use while adhering to a set of safety standards;
- Sending these steps to the data center and putting them into action;
- The local control system verifies the acts.

KEYWORDS

- **dynamic programming**
- **Internet of things**
- **light detection and ranging**
- **local control system**
- **South Eastern Pennsylvania Transportation Authority**
- **virtual private network**

REFERENCES

1. Bellemare, M. G., Dabney, W., & Munos, R., (2017). A distributional perspective on reinforcement learning. *Proceedings of the 34th International Conference on Machine Learning, in Proceedings of Machine Learning Research, 70*, 449–458.
2. Cobbe, K., Klimov, O., Hesse, C., Kim, T., & Schulman, J., (2019). Quantifying generalization in reinforcement learning. *Proceedings of the 36th International Conference on Machine Learning, Proceedings of Machine Learning Research, 97*, 1282–1289.
3. Matthew, B., Sam, R., Jane, X. W., Kurth-Nelson, Z., Charles, B., & Demis, H., (2019). Reinforcement leaning, fast and slow. *Trends in Cognitive Sciences, 23*(5), 408–422. ISSN 1364-6613. https://doi.org/10.1016/j.tics.2019.02.006.

4. Zhe, W., & Tianzhen, H., (2020). Reinforcement learning for building controls: The opportunities and challenges. *Applied Energy, 269*, 115036. ISSN 0306-2619. https://doi.org/10.1016/j.apenergy.2020.115036.
5. Okic, M., (2010). Adaptive ε-greedy exploration in reinforcement learning based on value differences. In: Dillmann, R., Beyerer, J., Hanebeck, U. D., & Schultz, T., (eds.), *KI 2010: Advances in Artificial Intelligence. KI 2010. Lecture Notes in Computer Science* (Vol. 6359). Springer, Berlin, Heidelberg. https://doi.org/10.1007/978-3-642-16111-7_23.
6. Hyeong, S. C., (2004). An ant system-based exploration-exploitation for reinforcement learning. In: *2004 IEEE International Conference on Systems, Man and Cybernetics (IEEE Cat. No.04CH37583)* (Vol. 4, pp. 3805–3810). doi: 10.1109/ICSMC.2004.1400937.
7. Dhiviya, S., Malathy, S., & Rajesh, K. D., (2018). Internet of things (IoT) elements, trends and applications. *International Journal on Computational and Theoretical Nanoscience, 15*(5), 1639–1643.
8. Peter, D., & Yael, N., (2008). Reinforcement learning: The good, the bad and the ugly. *Current Opinion in Neurobiology, 18*(2), 185–196. ISSN 0959-4388. https://doi.org/10.1016/j.conb.2008.08.003.
9. Vanitha, K., Vanitha, C. N., Mohamed, M. M., & Malathy, S., (2020). Efficient semantic interrogation scheme over cryptographic data in cloud. *IEEE Xplore Digital Library*. doi: 10.1109/ICICT48043.2020.9112383, June.
10. Zhenning, L., Hao, Y., Guohui, Z., Shangjia, D., & Cheng-Zhong, X., (2021). Network-wide traffic signal control optimization using a multi-agent deep reinforcement learning. *Transportation Research Part C: Emerging Technologies, 125*, 103059. ISSN 0968-090X. https://doi.org/10.1016/j.trc.2021.103059.
11. Manchella, K., Umrawal, A. K., & Aggarwal, V., (2021). FlexPool: A distributed model-free deep reinforcement learning algorithm for joint passengers and goods transportation. In: *IEEE Transactions on Intelligent Transportation Systems* (Vol. 22, No. 4, pp. 2035–2047). doi: 10.1109/TITS.2020.3048361.
12. Malathy, S., & Vanitha, C. N., (2019). Secure integration of cyber security and internet of things in addressing its challenges. *International Journal of Recent Technology and Engineering, 8*(4). ISSN: 2277-3878.
13. Nahid, P. F., Bo, Z., Tanvir, A., & Limon, B., (2021). Deep reinforcement learning in transportation research: A review. *Transportation Research Interdisciplinary Perspectives, 11*, 100425. ISSN 2590-1982. https://doi.org/10.1016/j.trip.2021.100425.

CHAPTER 8

Role of AI and IoT in Intelligent Transportation

PAWAN WHIG[1], ARUN VELU[2], RAHUL REDDY NADIKATTU[3], and YUSUF JIBRIN ALKALI[4]

[1]*Senior IEEE Member, Dean Research, Vivekananda Institute of Professional Studies, New Delhi, India*

[2]*Director Equifax and Researcher, Atlanta, USA*

[3]*Senior IEEE Member, Department of IT, University of Cumbersome, USA*

[4]*ICT Support Officer, Federal Inland Revenue Service, Nigeria*

ABSTRACT

The transport industry is being transformed by artificial intelligence (AI). It is now implemented in various transport areas, from helping vehicles, trains, ships, and airplanes to work autonomously, to make traffic flows smoother. In addition to making our lives better, it will help make other forms of travel easier, safer, more effective and smarter. AI-led autonomous transport could for instance help to reduce the human errors that are involved in many traffic accidents. These prospects, though, raise real obstacles, including unforeseen effects and misuse, such as cyber-attacks and skewed transport decisions. There are also job consequences and legal concerns related to responsibility for actions made on the spot by AI on humans. Society has accepted the concept of smart transportation and has implemented it broadly. At the same time, smart transportation

Artificial Intelligence for Future Intelligent Transportation: Smarter and Greener Infrastructure Design. Rajesh Kumar Dhanaraj, Nilayam Kumar Kamila, Subhendu Kumar Pani, Balamurugan Balusamy, & Vani Rajasekar (Eds.)

challenges are emerging, and a new path for its growth is given by the internet of things (IoT). The dream of a smart city administration structure using strong and sophisticated connectivity technology has been realized by IoT-based transport networks. In this chapter, we will concentrate on the smart transportation IoT system and innovations that play a key role in the management system of smart transport. And also, we will look at the role of IoT and AI in the transportation industry and highlight different applications where IoT architecture can be implemented.

8.1 INTRODUCTION

According to analysts, by 2020, 75% of new cars will have IoT connectivity [1]. Although the percentage rise refers to consumer applications, the concept of connected automobiles should pique the interest of other industries like shipping, logistics, and transportation [2]. Leaders in these fields would be prudent to prepare for a future in which AI and the internet of things (IoT) alter transportation management [3]. The era in which we live, the present, is the most technologically sophisticated in history [4]. Advancements in every business area, from customer-facing technology to predictive inventory to transportation management, are coming as a result of these new technologies [5].

Smart technology is changing the way businesses operate, and it's having an impact on your business whether you recognize it or not. Businesses are using AI and IoT at a breakneck pace to streamline processes and optimize data [6]. This technology has already made its mark in large cities, and it's swiftly gaining traction in transportation management.

The era we live in, the present, is the most technologically advanced in history. Advancements in every business area, from customer-facing technology to predictive inventory to transportation management, are coming as a result of these new technologies [7].

Smart technology is changing the way businesses operate, and it's having an impact on your business whether you recognize it or not [8]. Businesses are using AI and IoT at a breakneck pace to streamline processes and optimize data. This technology has already made its mark in large cities, and it's swiftly gaining traction in transportation management [9]. Transportation management, among all urban responsibilities, presents an intriguing problem, even for the most advanced AI tools and technologies.

City traffic is a highly dynamic environment in which thousands of participants using various modes of transportation interact in complex ways. Furthermore, decisions must be made in real time to ensure the safety and well-being of all traffic participants. In such an environment, activity planning is a difficult task [10]. Fortunately, AI-powered Smart City technologies are already making significant strides in addressing some of the most pressing transportation management challenges [11].

8.2 BACKGROUND

All shipments were handled by carriage and horses before Henry Ford developed the first car in 1908 as shown in Figure 8.1. Companies had to fight these methods with poor safety and efficiency. When automobiles began to appear on the markets and on roads, people began to use them for the delivery of goods [12]. The time spent shipping products and logistics costs have been reduced. Later in the end of the 20th century, a major change in logistics changed the delivery process: software logistics. Since then, technology has taken over multiple tasks, including paperwork and manual workflows, while securing information that is vulnerable [13].

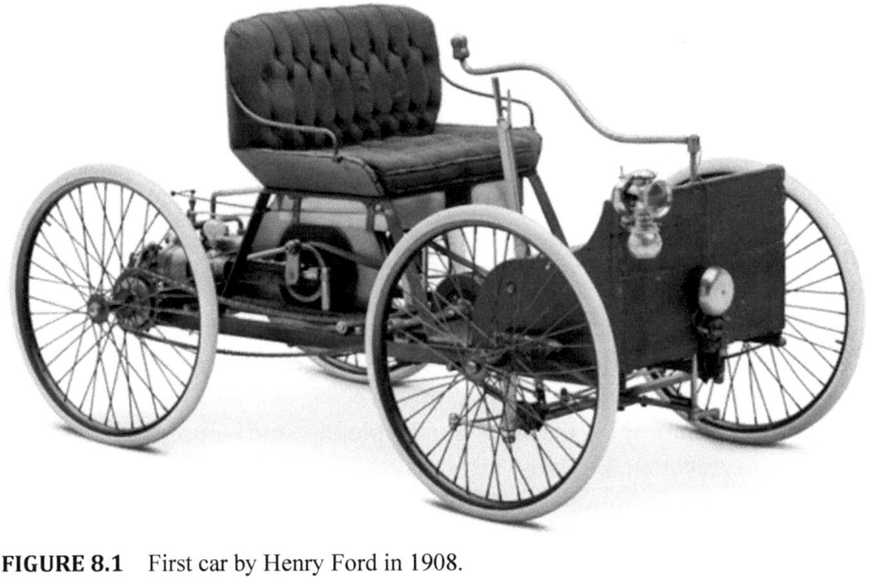

FIGURE 8.1 First car by Henry Ford in 1908.

Technology continually disrupts the logistics sector and alters the management of freight, sales, materials, goods, manufacture, and inventory [14]. Businesses always look for a solution that will provide information and help to reduce the high cost of logistics operations. As a consequence of the speed, intelligence, and efficacy, new technologies including AI, IoT, and blockchain have been adopted by the logistics sector in order to meet increasing demand and to address complex processes [15].

In this chapter, we will discuss the various ways of logistics and transportation development using latest technologies like artificial intelligence (AI) and IoT.

8.3 LOGISTICS AND TRANSPORTATION

The field of logistics and transport management requires analysis and accuracy. It monitors goods or materials delivery to customers from suppliers [16]. Logistics professionals are responsible for concentrating primarily on the procurement and planning of goods and materials for transport.

The logistics and transport management system mainly handles are follows and the main functions are also represented in Figure 8.2:

- Request for collection and delivery;
- Optimization of carrier management pickup;
- Management of warehouse;
- Delivery of transit;
- Reporting and BI.

8.4 REQUEST FOR COLLECTION AND DELIVERY

The client's role is to provide the carrier with pickup and delivery demands. This is how an interface can be developed for customers to send requests for collection and delivery [17].

A new solution to manage goods pickup and supply from scratch cannot be developed. Many logistics software, including NetSuite and McLeod, is available on the market and can be used to manage operations efficiently. The client's role is to provide the carrier with pickup and delivery demands [18].

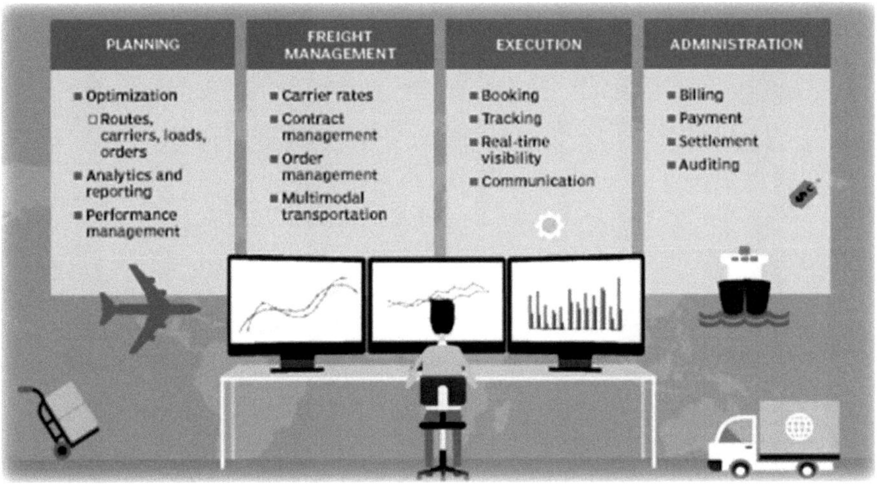

FIGURE 8.2 Main function of transportation management system.

8.4.1 NETSUITE

NetSuite is a mainstream coordination stage for the incorporated determining and planning, inventory network and stock, income the executives, client relations and business insight [19]. It is a product of Oracle and the UI Interface is look as given in Figure 8.3.

NetSuite is an online service that provides companies with the ability to manage every important business process in one single system [20]. There is no hardware, no high and early license fee, no hardware or software maintenance fees and no complex setups.

Companies use the NetSuite for corporate resource planning (ERPs) and inventory management, financial tracking, e-commerce shopping, and CRMs. This flexible platform is suitable for a variety of business applications [21].

In order to reduce their operating costs to increase their competitive advantage, many companies turn to the cloud to deliver their business systems. NetSuite, the Cloud ERP offers a set of cloud-based financial and enterprise resource planning (ERP) and ombudsman ship trading software, which operate in more than 200 countries and territories with over 18,000 customers [22].

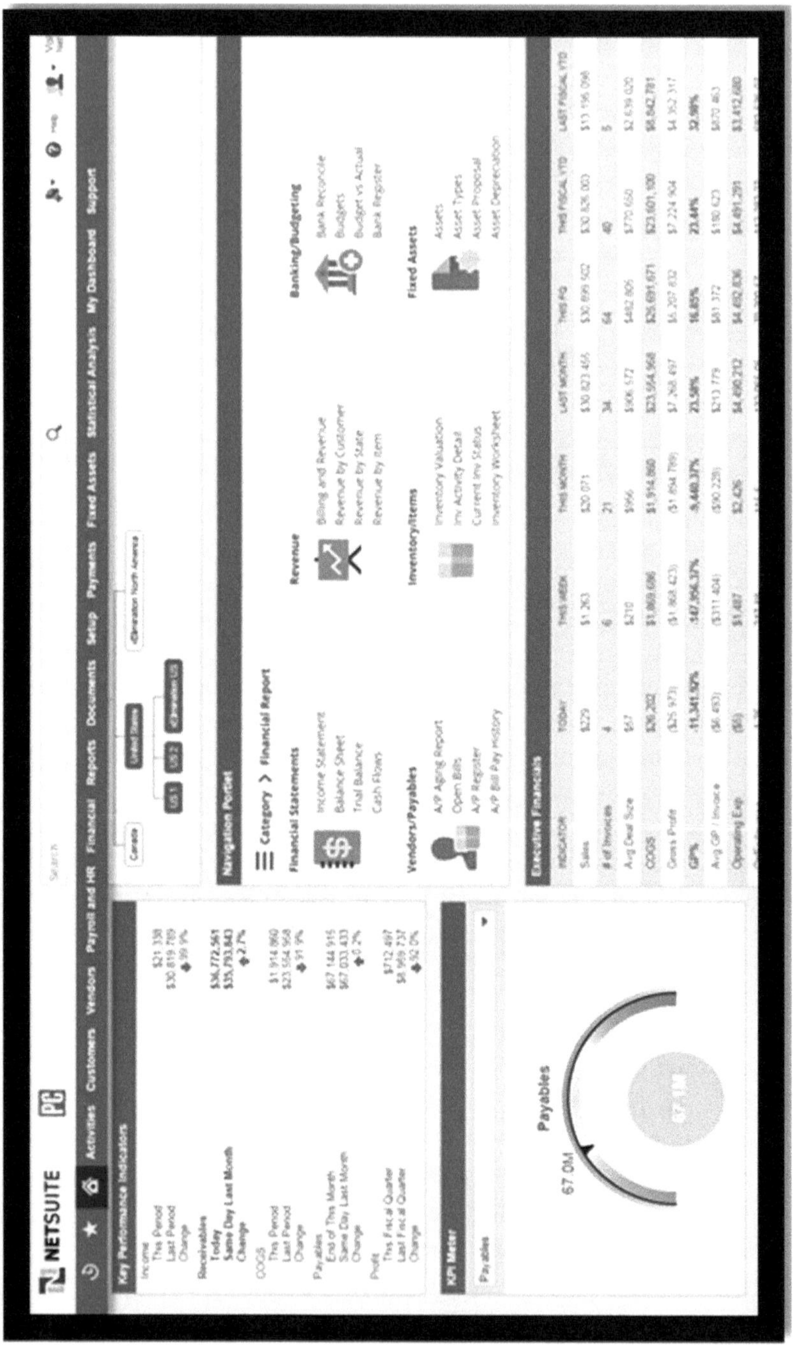

FIGURE 8.3 UI of NetSuite.

8.4.2 MCLEOD

McLeod provides lorry software and transport management solutions and manages lorry operations, document imaging, fleet management, automation processes and EDI. To manage pickup and delivery requests, one can use the solutions mentioned above the UI (user interface) of McLeod is shown in Figure 8.4. At the point when a client orders pickup or conveyance, the request is shipped off the dispatch, who allots the shipment to different drivers. Transporters should decide if the pickup or conveyance is done on time, just as how often the driver digresses from the booked course. It is feasible to do as such by joining AI and IoT into coordination applications [23].

McLeod Software is dedicated solely to the transportation industry with an established base of over 1,000 active customers in North America. This means that the needs and the complicated details of carriers, brokers, and freight management companies of all types are fully understood [23].

They also run regional offices with training facilities in Salt Lake City, Utah, and Downers Grove, Illinois, and own McLeod Software in Their national Headquarters in Birmingham, Alabama. They employ over 500 professionals, most of whom are developers of software [24].

McLeod Software plays a key role in the trucking and brokering industry. They are members and members of over 20 state and regional organizations, in addition to the national organizations ATA, TCA, TIA, AFTC, NTTC, and NPTC [25–27].

In the supply chain operations, Gartner has predicted the use of advanced analysis, IoT, and AI by up to 50% of major global companies [28].

8.5 OPTIMIZATION OF CARRIER MANAGEMENT PICKUP

Carriers can identify whether deliveries are efficient or not using predictive analytical algorithms. Different Types of analytics is summarized in Figure 8.5.

8.5.1 PREDICTIVE ANALYTICS

Effective predictive logistics analytics ensure fleets are received and shifted in due course, to provide customers with shipments as they wish [29].

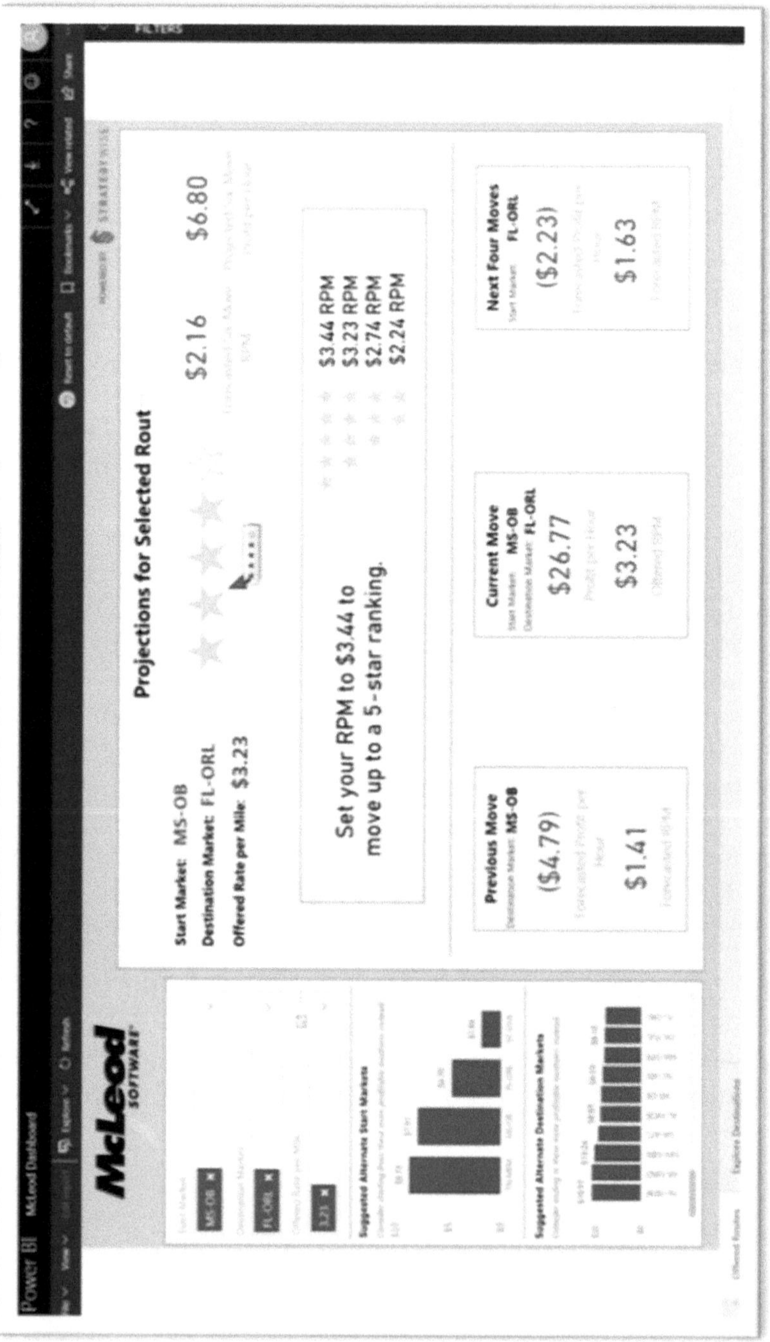

FIGURE 8.4 UI of McLeod.

Inserted in trucks, prepares or delivers feed information like motor execution and speed, sensor-enacted resources or IoT gadgets send and model the assessed appearance times and motor flaws to the transporters. For example, telematic information from a vehicle can show its excess speed, position, condition, and time to arrive at the objective [30]. The recorded information can be utilized to advise beneficiaries of deferrals alongside load-down, lorries headed towards similar objectives and necessities for item conveyance. This guarantees that postponements and client assumptions are limited [31].

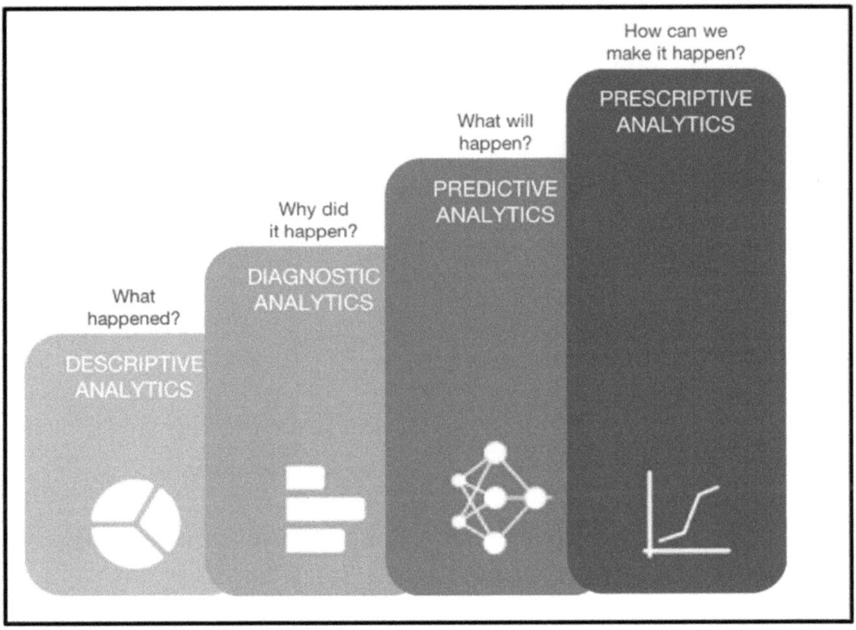

FIGURE 8.5 Different types of analytics.

Intelligent AI and IoT logistics help ports, shippers, vendors, and agents optimize the use and schedules of resource.

8.5.2 FORECASTING ON THE BASIS OF INVENTORY

Companies can anticipate demand for shipments and goods throughout the supply chain with logistics demand forecasts. In order to evaluate the demand of capacity based on historical data, including inventory data and order data, logistics companies need to set up a predictor model [32].

The basic practice for forecasting based on Inventory is shown in Figure 8.6.

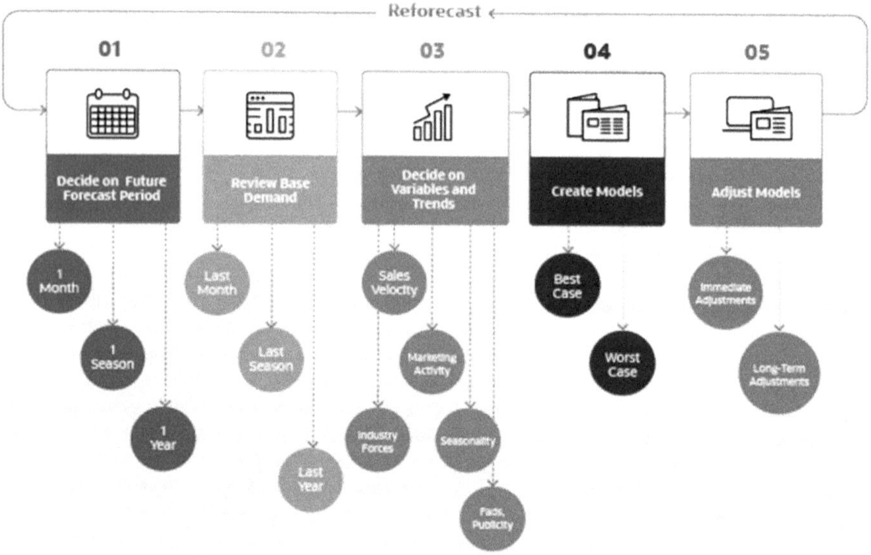

FIGURE 8.6 Basic practice for forecasting based on inventory.

Source: Reprinted from https://www.netsuite.com.au/portal/au/resource/articles/inventory-management/inventory-forecasting.shtml

With custom interest determining models, organizations can accomplish a precise estimate to assist them with understanding the extra limit required, lessen kilometers of repositioning resources and improve the limit and usage of freight vehicles [33].

Since coordination ventures can precisely conjecture the requests for resources and shipments, they can build load limit and give clients popularity merchandise on schedule.

8.5.3 IOT AND ARTIFICIAL INTELLIGENCE (AI)-DRIVEN

The application of IoT logistics sensors makes goods tracking more accessible. Data collection and exchange is done using sensors. IoT enables data handling over the network infrastructure remotely [34]. The basic structure is given in Figure 8.7.

In carriers carrying goods sensors are integrated. Sensor captured data can be converted to use of AI to provide valuable insight.

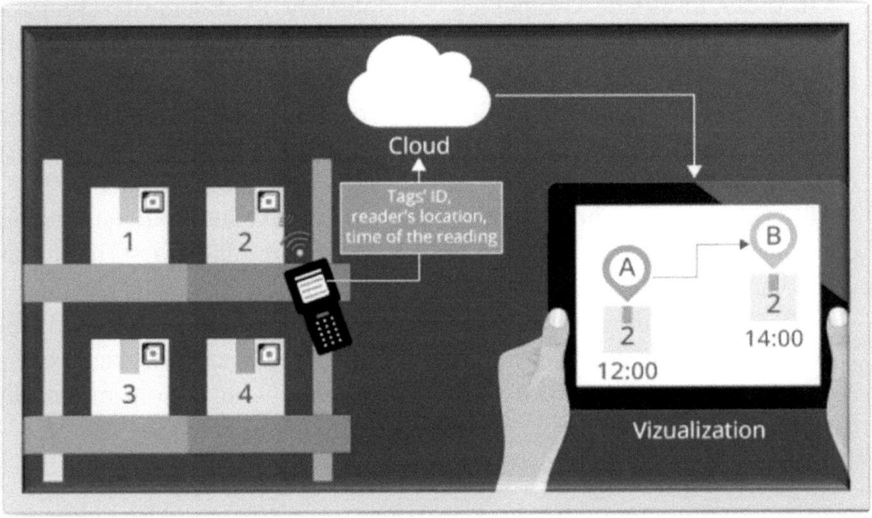

FIGURE 8.7 Basic structure of IoT and artificial intelligence driven.

AI empowered investigation empower shipments to be followed from flight focuses to the objective and send continuous following reports [35]. Effective product monitoring provides information, including:

- Manipulation or robbery while traveling;
- Time of departure and arrival;
- Live shipping location;
- Any differences from the planned route.

Logistics companies are able to improve the collection and delivery of orders with all the information till the delivery.

Future interest that helps item advancement and organic market arranging can be considered typical from the information gathered during the store network. The manufacturing companies can continue optimizing the production of goods when the product is reached at the final destination.

Take the example of foodstuffs processed. The food production status, temperature at which it is stored, and time of delivery can be monitored using different types of sensors. Analysis of hazards and criticisms in order to ensure that food quality isn't hindered, checklists are used throughout production and delivery procedures [36].

In order to enable companies to implement food security solutions, IoT sensors send useful information relating to food in the supply chain. It contributes not only to complying with rules on the safety of food and also to preserving customer faithfulness and confidence by total clearness.

The carrier's role is also to provide the optimized routes for the organized delivery of the goods and interact with warehouse owners [37].

Improvement by shrewd advances, including AI and IOT, can upgrade coordination and decrease the time needed for products assortment and conveyance.

IoT provides information that can be monitored and reported in real time. It is much easier to determine whether delays arise because of bad weather conditions or problems with truck maintenance with remote monitoring capabilities. Thus, the supply chain can effectively monitor the carriers.

8.5.3.1 ROLE OF AI/IOT IN PLANNING

Loads can be allocated to drivers as per their flow area for pickup. Work to pick up goods are assigned to drivers next to the collection site and drivers are informed of the optimized route towards the warehouse [38].

For exchange or delivery, trucks and warehouses shall be selected on the basis of, for example, the type of goods to be collected.

IoT sensor-equipped trucks deliver real-time route optimization analysis, reliability, and time reduction. The logistics service providers receive data and alerts captured by the sensors.

IoT sensor information includes:

- Goods type;
- The carrier's real-time location;
- Abweichungen from the intended route;
- Warehouses pick-up/pick-up;
- Humidity/temperature.

Top five route planning startups impacting logistics and fleet management is shown in Figure 8.8.

Man-made intelligence on the information gathered can assist you with anticipating the accompanying elements:

- Probable time of supply;
- Investigation of driver conduct;
- Quality of products such as perishable products;
- Differences between route planned and route actual.

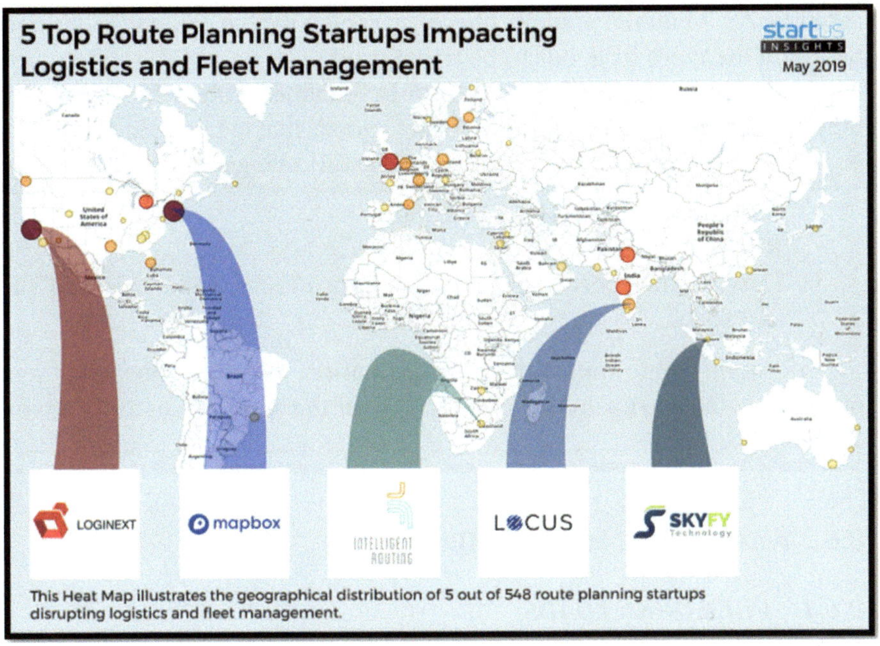

FIGURE 8.8 Top five route planning startups.

Source: Reprinted from StartUs Magazine, May 2019. https://www.startus-insights.com/innovators-guide/5-top-route-planning-startups-out-of-548-in-logistics/

Information assortment and investigation would help coordination organizations decrease superfluous expenses, improve an opportunity to gather and convey merchandise and give upgraded streets. It is feasible to see how the real course varies from the arranged course and why deviations from the arranged course may happen. This combined AI and IoT projections for future delivery and pick-up routes more optimized.

Transit and exchange can be planned with optimized route planning. Since distribution centers can acquire data about the assessed season of appearance of the drivers, they can design the vehicle of products effectively. For instance, devices and work needed to dump products can be orchestrated already to empower speedy travel/trade.

The pick-up of orders must be optimized in a way that products are not in any way affected. The optimized pick-up of technologies such as AI and IoT can be guaranteed.

Trucks for pick-up are selected according to the load type to be transported. AI lets carriers accurately take decisions and learn historical patterns. An AI model predicts the correct truck for the collection of goods based on the types of goods to be transferred [39].

One can plan pickup of goods into IoT-enabled trucks, for instance, if they have to collect perishable or sensitive goods that require additional care.

The trucks enabled by IoT can collect the following product information:

- Temperature of stocking of goods;
- Real time location;
- Data on transport truck moisture exposure, including cost of fuel and speed.

Information is imparted to coordination organizations and item proprietors to keep up quality command over the whole inventory network cycle.

8.6 WAREHOUSE MANAGEMENT

8.6.1 PREPARING LOADS

By connecting AI and IoT together, carriers, and warehouses can enhance collaboration. For instance, in a truck equipped IoT sensors send warehouse managers the real-time location. Manager can maintain vacant space and prepare for unloading products before time by using the information provided by sensors [40]. It helps warehouses to efficiently and accurately manage their schedules. Intelligent prediction using AI can be used in time for preparation of load and unload. As a result, both logistics companies and warehouse operators have less waiting time. The warehouse management flowchart is also shown in Figure 8.9.

8.6.2 PLANNING OF DOORS

Sensors introduced around the store region would send entryway related data to lorry drivers based on their GPS information and their assessed appearance time. It helps transporters to decrease the time they need to track down the correct way to enter the stockroom rapidly. AI and IoT therefore also make it easy to plan a free door between logistical companies and warehouses to facilitate better collaboration.

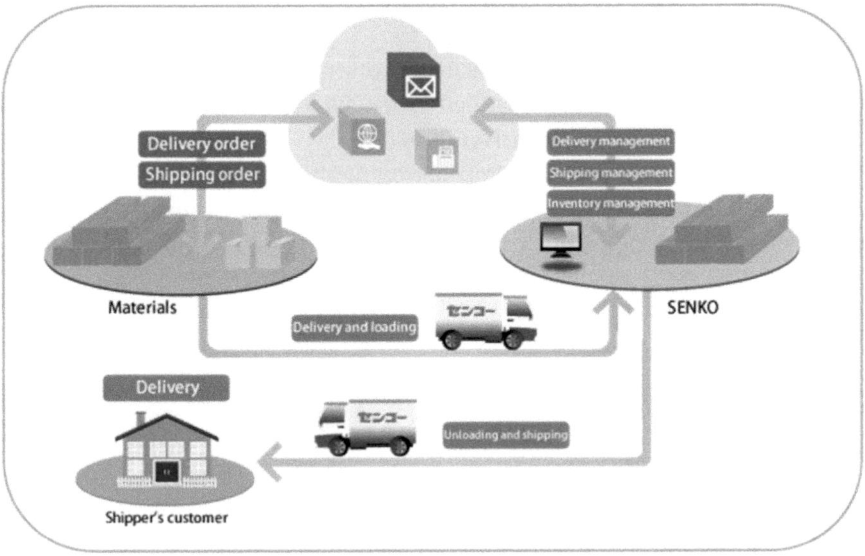

FIGURE 8.9 Warehouse management flowchart.

8.6.3 *COMPLIANCE AND SECURITY*

The unloading/loading of goods from the truck must be avoided. It is essential. To provide security, you can use IoT-enabled locks that allow air brakes to be interlocked with the door of the trailer. Clever locks make it impossible for the truck to leave until unloading is completed. Only when the dock door is closed can trailers leave the warehouse. It can safely maintain your equipment and staff.

The utilization of PC screen is another approach to empower stockroom security. You can introduce IoT cameras in the heap/dump status and burden move stockroom territory. Edge figuring can be applied to shrewd cameras which permit the critical activity dependent on any occasion found during load move tasks.

In the event that any delicate thing separates during dumping, for instance, because of terrible work conduct, stockroom supervisors can be educated about it.

8.6.4 DELIVERY OF TRANSIT

It is possible to transform transit operations using AI and IoT-enabled sensors. The GPS sensors installed in trucks are based on the AI model's real-time assessment of the arrival time.

IoT sensors outfitted with a truck can gather data, including episodes of crash, temperature, and stickiness of products. Logistics companies can track transit operations in real time using real-time data collected through IoT devices. The flowchart of Delivery of transit is shown in Figure 8.10.

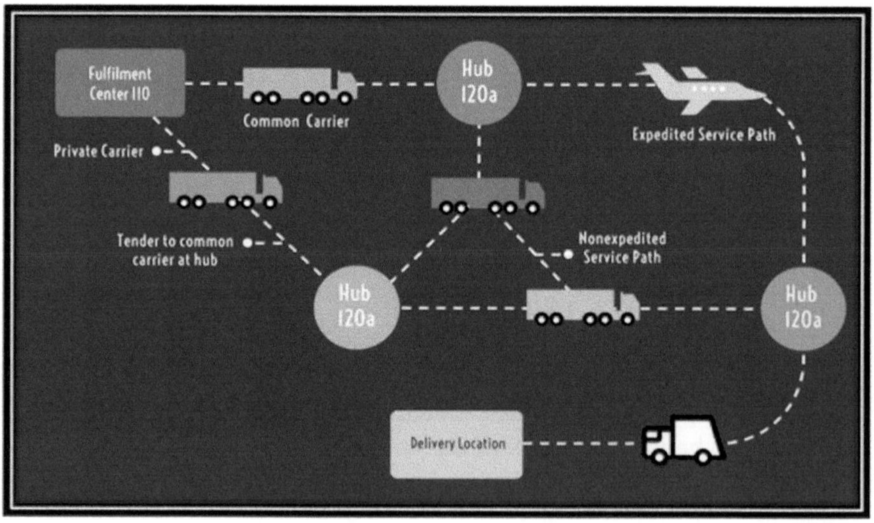

FIGURE 8.10 Flowchart of delivery of transit.

Implementing predictive analytics would help companies understand how long the truck will have to reach the destination, depending on the predicted time. If the goods are delivered at the right temperature conditions, customers, and logistics companies can know.

8.6.5 OPTIMIZATION OF LOAD EXCHANGE

Should the truck be over-burden, or a mishap happen, transporters should stand by quite a while to look for options. Notwithstanding, IoT and AI can assist with decreasing holding up time by catching continuous information and permitting keen activities. AI based models can be executed to advance the heap trade starting with one truck and then onto the next.

For instance, if a lorry meets a mishap, IoT sensors will gather this data and another vehicle is consequently appointed to supplant the heap and send it to the last objective. The truck's GPS provide details such as latitude and longitude, the location of the truck in real time and the movement of the truck. The carriers can therefore quickly determine a vehicle's condition during an accident or misapprehension and exchange arrangements can be taken accordingly.

In this manner, new advancements guarantee the smooth travel and supply of products. The subsequent stage is to supply the merchandise to end-clients. The transporter was going to the stockroom to supply merchandise to the last clients. Like the above-mentioned, carriages would be told of the heap arrangement of the empty entryway for pickup of products and stockroom administrators.

Under full security, the load will be transferred to the carrier using IoT-enabled locks and computer vision cameras. The carrier will depart once loading has been completed and the goods delivered to the clients.

8.7 REPORTING AND BI

Whenever items are gathered and conveyed, organizations would have a complete report, which would incorporate the negative and positive execution patterns in the conveyance of products.

As companies need granular transparency for their transport expenses in order to effectively manage and control them, the demand for business intelligence within logistics and transport is on the rise (Figure 8.11).

Business intelligence enables data to be converted into valuable data. In the past, the reporting was restricted to data extraction, collection, and transfer from a system in a tablet or database.

But the next level has now been reached by business intelligence where companies are able to produce valuable reports which display all logistics providers data in a scorecard format. The criteria for determining carrier performance are assigned to factors such as pickup and delivery on time, capacity commitments, and driver behavior.

The board likewise can utilize continuous dashboards that give ongoing data and assist clients with taking care of issues when they emerge. They require a day by day and fast outline of what's going on. Dashboards permit organizations to profit by fast response times since clients don't need to stand by until somebody makes reports and sends them.

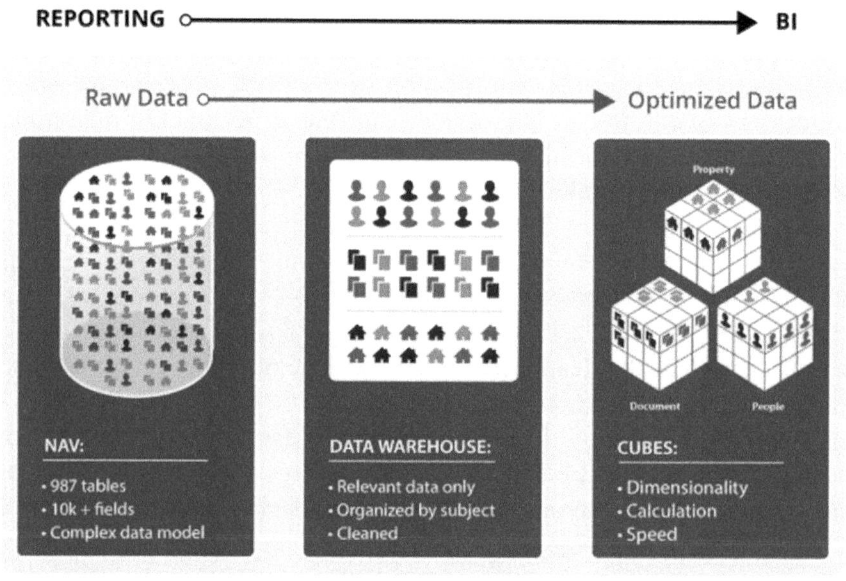

FIGURE 8.11 Example to show business intelligence and reporting.

Source: Reprinted from https://insightsoftware.com/blog/business-intelligence-vs-reporting-finding-your-bread-and-butter/

Information that can foresee openings and future dangers on the inventory network/transport network is utilized by organizations as BI and announcing.

Let's see how suppliers can take an example to improve logistics and supply chain operations through business information.

Let that a shipper has an 89% charge transfer on delivery time, but wishes to attract another 10% of the shipment at the root of the problem. Due to the broken lanes did the delays occur? Does the equipment or carrier have any problems?

It is possible to limit delay factors with business intelligence. Let, a shipper reaches late because of a reason, business intelligence can discover the problem and take action.

BI data facilitates the efficiency of operational decision making for logistics companies and their team. The implementation of BI across shipments can thus lead to end-to-end pick-up and delivery times.

8.8 SUMMARIZE

Man-made reasoning and the IoT disturb coordination and transport by making coordination tasks step by step more brilliant. Pick-up/delivery delays and lorry capacity problems are expected to become part of the past when AI and IoT are presented.

KEYWORDS

- **artificial intelligence**
- **business intelligence**
- **enterprise resource planning**
- **innovation**
- **Internet of things**
- **transportation**

REFERENCES

1. Sadek, A., (2007). Artificial intelligence in transportation. *Transp. Res. Circ., E-C113*, 72–79.
2. Yegnanarayana, B., (1999). *Artificial Neural Networks* (p. 476). PHI Learning Pvt. Ltd.: New Delhi, India.
3. Abraham, A., (2005). Artificial neural networks. In: Sydenham, P. H., & Thorn, R., (eds.), *Handbook of Measuring System Design*. John Wiley & Sons, Ltd.: Hoboken, NJ, USA.
4. Minsky, M., & Papert, S., (1969). *Perceptron Expanded Edition*. MIT Press: Cambridge, MA, USA.

5. Wolpert, D. H., (1993). Bayesian backpropagation over I-O functions rather than weights. *Adv. Neural Inf. Process. Syst., 6*, 200–207.
6. Parker, D., (1985). *Learning Logic: Technical Report TR-87, Center for Computational Research in Economics and Management Science.* The MIT Press: Cambridge, MA, USA.
7. Getoor, B., & Taskar, L., (2007). *Introduction to Statistical Relational Learning; Volume L of Adaptive Computation and Machine Learning.* MIT Press: Cambridge, MA, USA.
8. Bacciu, D., Micheli, A., & Sperduti, A. (2012). "Compositional Generative Mapping for Tree-Structured Data—Part I: Bottom-Up Probabilistic Modeling of Trees." *IEEE Transactions on Neural Networks and Learning Systems, 23*(12), 1987–2002, doi: 10.1109/TNNLS.2012.2222044.
9. Bacciu, D., Micheli, A., & Sperduti, A. (2012). Compositional generative mapping for tree-structured data–part I: bottom-up probabilistic modeling of trees. *IEEE Trans Neural Netw Learn Syst. 23*(12), 1987–2002. doi: 10.1109/TNNLS.2012.2222044. PMID: 24808152.
10. Krizhevsky, A., Sutskever, I., & Hinton, G. E., (2012). ImageNet classification with deep convolutional neural networks. *Adv. Neural Inf. Process. Syst., 60*, 1–9.
11. Kim, P., (2017). *Convolutional Neural Network.* MATLAB Deep Learning A press: Berkeley, CA, USA.
12. McCann, M. T., Jin, K. H., & Unser, M., (2017). Deep convolutional neural network for inverse problems in imaging. *IEEE Signal Process. Mag., 34*, 85–95.
13. Pawan, W., & Ahmad, S. N., (2014). CMOS integrated VDBA-ISFET device for water quality monitoring. *International Journal of Intelligent Engineering and Systems* (Vol. 7, No.1). Accepted for publication. (Scopus) ISSN: 2185-3118.
14. Pawan, W., & Vaibhav, B., (2014). Performance analysis of multi-functional bot system design using microcontroller. *International Journal of Intelligent Systems and Applications, 02,* 69–75. ISSN No: 2074-9058.
15. Pawan, W., & Ahmad, S. N., (2014). Development of low power dynamic threshold PCS system. *Journal of Electrical and Electronic Systems, 3*(3), 1–6. ISSN No: 2332-0796.
16. Pawan, W., & Ahmad, S. N., (2015). Novel FGMOS based PCS device for low power applications. *Photonic Sensor (Springer), 5*(2), 1–5. (SCI, ISI Index) ISSN No: 1674-9251.
17. Pawan, W., & Ahmad, S. N., (2015). Impact of parameters on characteristic of novel PCS. *Canadian Journal of Basic and Applied Science, 3*(2), 45–52. ISSN No: 2292-3381.
18. Ruchin, C. M., & Pawan, W., (2015). Design and simulation of dynamic UART using scan path technique (USPT). *International Journal of Electrical, Electronics & Computer Science Engineering, 1*, 6–11. ISSN No: 2348-2273.
19. Zhang, X. Y., Yin, F., Zhang, Y. M., Liu, C. L., & Bengio, Y., (2018). Drawing and recognizing Chinese characters with recurrent neural network. *IEEE Trans. Pattern Anal. Mach. Intell., 40*, 849–862. [CrossRef] [PubMed].
20. Du, M. N. S., & Swamy, K. L., (2014). Recurrent neural networks. In: *Neural Networks and Statistical Learning.* Springer: London, UK.

21. Caterini, D. E., & Chang, A. L., (2018). Recurrent neural networks. In: *Deep Neural Networks in a Mathematical Framework*. Springer: Cham, Switzerland.
22. Patterson, D., (1990). *Introduction to Artificial Intelligence and Expert Systems*. Prentice-Hall, Inc.: Upper Saddle River, NJ, USA.
23. Australian Infrastructure Audit, (2015). *Infrastructure Australia*. Our Infrastructure Challenge.
24. Linking Melbourne Authority. *Linking Melbourne*. Annual Report. Available online: https://www.parliament.vic.gov.au/file_uploads/Linking_Melbourne_Authority_Annual_Report_2014-2015_CwBGv8WN.pdf (accessed on 1 January 2023).
25. Klügl, F., Bazzan, A. L. C., & Ossowski, S., (2010). Agents in traffic and transportation. *Transp. Res. Part C Emerg. Technol., 18*, 69–70.
26. Dogan, E., & Akgüngör, A. P., (2013). Forecasting highway casualties under the effect of railway development policy in Turkey using artificial neural networks. *Neural Comput. Appl., 22*, 869–877.
27. Budalakoti, S., Srivastava, A. N., & Otey, M. E., (2009). Anomaly detection and diagnosis algorithms for discrete symbol sequences with applications to airline safety. *IEEE Trans. Syst. Man Cybern. Part C Appl. Rev., 39*, 101–113.
28. Akgüngör, A. P., & Dogan, E., (2009). An artificial intelligent approach to traffic accident estimation: Model development and application. *Transport, 24*, 135–142.
29. Dia, H., & Rose, G., (1997). Development and evaluation of neural network freeway incident detection models using field data. *Transp. Res. Part C Emerg. Technol., 5*, 313–331.
30. Wang, R., Fan, S., & Work, D. B., (2016). Efficient multiple model particle filtering for joint traffic state estimation and incident detection. *Transp. Res. Part C Emerg. Technol., 71*, 521–537.
31. Wang, R., & Work, D. B., (2014). Interactive multiple model ensemble Kalman filter for traffic estimation and incident detection. In: *Proceedings of the 17th International IEEE Conference on Intelligent Transportation Systems (ITSC)* (pp. 804–809). Qingdao, China.
32. Dia, H., (2001). An object-oriented neural network approach to short-term traffic forecasting. *Eur. J. Oper. Res., 131*, 253–261.
33. Huang, W., Song, G., Hong, H., & Xie, K., (2014). Deep architecture for traffic flow prediction: Deep belief networks with multitask learning. *IEEE Trans. Intell. Transp. Syst., 15*, 2191–2201.
34. Pawan, W., & Ahmad, S. N., (2012). Performance analysis of various readout circuits for monitoring quality of water using analog integrated circuits. *International Journal of Intelligent Systems and Applications (IJISA)* (Vol. 4, No.11, pp. 91–98). ISSN: 2074-904X (Print), ISSN: 2074-9058 (Online).
35. Pawan, W., & Ahmad, S. N., (2013). A novel pseudo PMOS Integrated CC-ISFET device for water quality monitoring. *Journal of Integrated Circuit and System* (Vol. 8, No. 2, pp. 1–6). ISSN, 1807-1953 (Scopus).
36. Pawan, W., & Ahmad, S. N., (2014). Simulation of linear dynamic macro model of photo catalytic sensor in spice. *COMPEL – the International Journal of Computation and Mathematics in Electrical and Electronic Engineering* (Vol. 33 No. 1, 2). ISSN: 0332-1649 (SCI, ISI index).

37. Vaibhav, B., & Pawan, W., (2013). A secured dual tune multi frequency based smart elevator control system. *International Journal of Research in Engineering and Advanced Technology, 4*(1). ISSN (Online): 2319-1163.
38. Pawan, W., & Ahmad, S. N., (2013). A novel pseudo NMOS integrated ISFET device for water quality monitoring. *Active and Passive Components* (Vol. 1, No. 1). Hindawi article I.D 258970. (Scopus). ISSN 0882-7516.
39. Vaibhav, B., & Pawan, W., (2013). Modeling and simulation of electrical load control system using RF technology. *International Journal of Multidisplinary Science and Engineering, 4*(2), 44–47. ISSN 2045-7057.
40. Pawan, W., & Ahmad, S. N., (2014). Development of economical ASIC for PCS for water quality monitoring. *Journal of Circuit System and Computers, 23*(6), 1–13.

CHAPTER 9

Potential Applications of AI in Transportation

C. N. VANITHA[1], S. MALATHY[2], C. SHARMILA[2], and T. C. KALAISELVI[3]

[1]*Professor, Department of Computer Science and Engineering, Kongu Engineering College, Perundurai, Tamil Nadu, India*

[2]*Assistant Professor, Department of Computer Science and Engineering, Kongu Engineering College, Perundurai, Tamil Nadu, India*

[3]*Professor, Department of Electronics and communication and Engineering, Kongu Engineering College, Perundurai, Tamil Nadu, India*

ABSTRACT

Over the past few hundred years, the transportation sector has witnessed numerous transformations and revolutions, and we are now at a point where significant advancements in the form of artificial intelligence are being made. AI is capturing the attention of transportation executives all around the world, whether it is through self-driving cars for greater dependability, road condition monitoring for increased safety, or traffic flow analysis for greater efficiency. AI is now more available than ever. The whole transportation sector is starting to incorporate AI into its offerings, including automakers, parking garages, cruise lines, and more. In this chapter, we'll look at a few more instances of artificial intelligence in the field of transportation and how it's addressing some of the most prevalent and enduring problems in this field.

Artificial Intelligence for Future Intelligent Transportation: Smarter and Greener Infrastructure Design. Rajesh Kumar Dhanaraj, Nilayam Kumar Kamila, Subhendu Kumar Pani, Balamurugan Balusamy, & Vani Rajasekar (Eds.)

9.1 INTRODUCTION

Artificial intelligence (AI) is transforming every aspect of life by allowing robots to learn from their mistakes, adapt, and complete tasks that previously required human cognition. The US government has designated AI as one of its top science and technology priorities. In Figure 9.1. The intelligent transportation systems (ITS) Joint Program Office (JPO) designated AI research as a priority area in order to hasten state and municipal agencies implementation of AI to solve transportation issues [1]. The objective of this chapter, which is based on a literature analysis, is to identify broad categories of AI-enabled applications that will be used to solve specific transportation challenges and requirements, as well as to summarize existing and potential AI-enabled applications within each category. Time-consuming and repetitive tasks are mechanized by AI-powered systems, which learn over time and eventually conduct key functions and make decisions on their own. Because of this one-of-a-kind potential, transportation firms began investing in AI technology to boost profits and stay ahead of the competition. Although the transportation sector has just lately begun to employ AI in essential tasks, transportation reliability and safety remain a concern. Safety, capacity, pollution, and reliability are all major transportation problems with a lot of room for AI innovation. Despite the challenges, AI is anticipated to reach 3.5 billion dollars in the transportation sector by 2023.

Transportation AI has progressed and is now being used in a variety of ways. In this study report, there are three major examples:

- In economic decision, marketing, and governance, AI is being applied. To address the issue of increased demands and high road availability, these are required. To better estimate travel times, traffic patterns, and incidents, more intended to reflect and line detection are used.
- The use of AI to improve public transportation is also highlighted. Because public transportation is considered a long-term means of mobility.
- Self-driving cars are now the potential AI use in transportation, with the goal of increasing productive capacity by decreasing the quantity of injury and death.

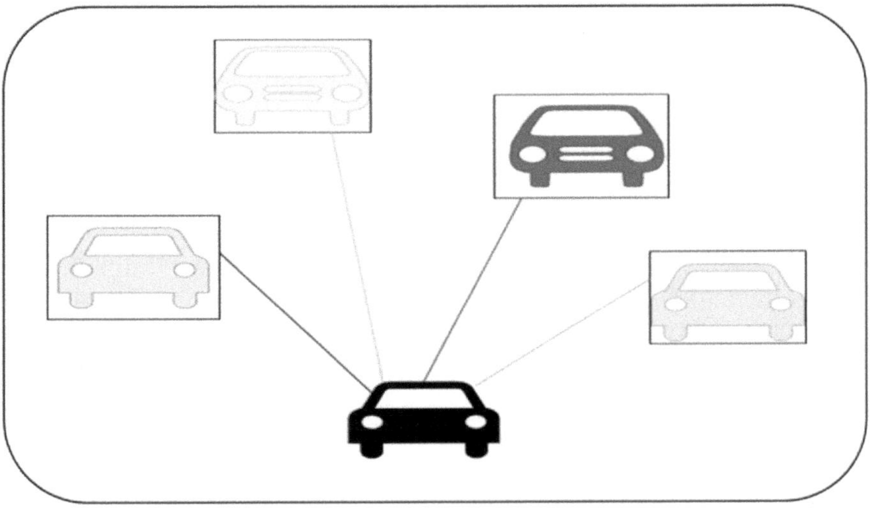

FIGURE 9.1 AI in transportation.

9.2 TRANSPORTATION NETWORK STRUCTURES IN AI: PLANNING, DESIGNING, AND CONTROLLING

The network design problem (NDP) involves creating an effective road strategy for transportation planning. If the capacity of present infrastructure changes, it can become a continuous difficulty, a significant problem if extra infrastructure is added, or a mix of continuous and discrete challenges if new infrastructure is introduced. In the 1990s, NNs for transportation planning, design, and simulation had been the subject of research.

A concurrent neural net approach was also used to evaluate the geometric distribution among transport and land planning. Most studies by environment 2019, 11 out of 24 employ Robotic to greater effect among the information as a vast amount of data with powerful algorithms becomes the focus of most studies. This has been providing a projected boost for urban expansion in China based on simulated economic land usage and land cover for many years. It focuses on developing a transportation system management and safety strategy that balances mobility needs with appropriate roadway, railway, and airway distribution [2]. They utilized ANN to predict accidents and injuries on a heavily used road between Istanbul and Ankara. In addition, vehicle route planning is necessary to avoid traffic and travel time delays. Many of them came to the conclusion that the ant colony algorithm is one of the most effective ways to solve the vehicle routing problem. While concentrating on applying the BCO algorithm to solve a route and wavelength challenge.

The task comprises picking a path inside an intranet and allocating a wavelength to that path for each linked terminal in order to increase the number of links between the devices in the system. They also focus on modeling and recognizing security infractions utilizing microscopic traffic data, as well as traffic control systems and highway management strategies. They develop and improve the real-time route generating system that relies on traveler choices and determine the optimal routes for public transit customers.

A machine learning (ML) system has also been created to provide signal processing and quick computing analytics capabilities to ITS devices. As IT evolves, data complexity will rise, necessitating the use of ML techniques to discover patterns and characteristics in this data in order to create a more integrated transportation system. At crossroads, for example, a ML model and fuzzy methodologies were utilized to control traffic light systems manually.

The proposed ITS solution, named "Never Stop," made use of RFID chips and was found to reduce average vehicle wait times. In addition, two NNs systems based on tiny, simulated data were built to better control the road. The first module contains traffic lights, while the second forecasts traffic congestion in the future. It showed an inter NNs system that was evaluated in several interconnection networks to show how Hidden layers may be used to regulate traffic. ANNs can also help with signal traffic control. Based on tiny, simulated data, they built two NNs algorithms to better control the road. AI is an ever-changing study field, with new applications published on a regular basis to take advantage of AI's skills to improve planning, decision-making, and leadership.

9.3 TRANSPORTATION AND ARTIFICIAL INTELLIGENCE (AI)

Already AI is being utilized in the transportation industry. As AI develops and matures, the variety of jobs in which AI is a significant part unquestionably increase. AI is commonly used in transportation in different ways which are discussed in subsections.

9.3.1 AUTONOMOUS VEHICLES (AVS)

Within the transportation industry, autonomous cars are now a reality. They are the first step toward a replacement mode of autonomous transportation, which previously appeared like a far-fetched concept [3]. These self-driving cars are powered by AI's processing, control, and optimization capabilities. Data transmission and processing in real time are critical in autonomous cars, and any ambiguity in these procedures might result in a catastrophic conclusion. Self-driving taxis have started operating in China, but the operator remains in the vehicle for safety purposes, ready to take operation of the vehicle in case of emergency. In the year 2018, the United States began to employ self-driving trucks. With trucks transporting 65% of goods worldwide. Autonomous transportation is also expected to save about 45% on maintenance and administration expenditures. Waymo has started testing self-driving minivans and trucks on public roads in a limited number of states. These types of initiatives are currently in the early stages of development, with the goal of developing autonomous cars that make passengers safe. Technology advances, driverless vehicles will become more reliable and widespread.

A global study was conducted on one aspect of what makes a cognitive autonomous car desirable to passengers. These were the following aspects:

- Self-healing;
- Self-socializing;
- Self-learning;
- Self-driving;
- Self-configuring;
- Self-Integrating.

9.3.2 PUBLIC TRANSPORTATION

The safety of passengers, workers, and drivers has long been a priority in the transportation industry. AI's application was the autonomous system, which is currently present in nearly every commercial jet and is an essential component of air travel [4]. According to the New York Times, just 7 mins of a normal Boeing aircraft are directed by pilots (preflight checks and arrivals), with the remainder handled by an autopilot. Since there are currently autonomous buses that transport people using sensors, cameras, and GPS, AI will have a big influence on public transportation is denoted in Figure 9.2.

FIGURE 9.2 Autonomous buses.

AI technology may help us in a variety of ways, including reducing human mistakes in traffic, monitoring compliance with safety rules, reviewing maintenance logs for vehicles and managing transportation in such cases a manner through which lessen the impact of traffic risks in the metropolitan area.

9.3.3 APPS FOR SMARTPHONES

Today, there's an app for that, and one of them is AI-powered actual traffic data via tools such as (Figure 9.3) google Navigation or Radar [5]. These apps make use of smartphone info on location to analyze traffic conditions in your neighborhood and give users with better traffic information. Why would you need an app if the automobile can do everything for you? These applications may soon encounter competition in the field of self-driving cars.

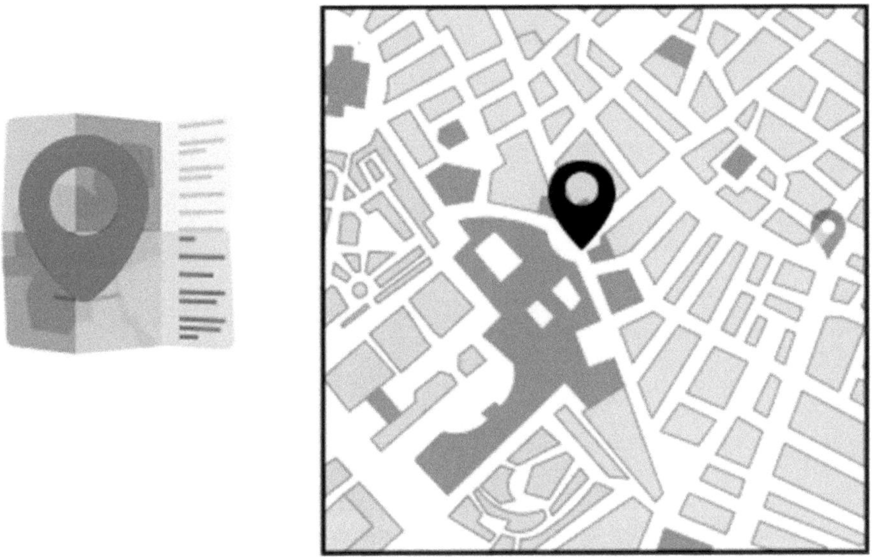

FIGURE 9.3 Google navigation.

9.3.4 TRAFFIC MANAGEMENT

Because of its capacities for processing, control, and optimization, AI May be used to manage transportation systems [6]. AI can be used in traffic management to make roads smarter and traffic flow more smoothly. Another significant benefit is the capacity to foresee physical concerns. This might result in excessive traffic or congestion (Figure 9.4).

Sensors and cameras installed on the roadways capture (Figure 9.4) a huge quantity of data on traffic patterns. Big data and AI technologies analyze this data to uncover traffic trends. The pertinent insights provide commuting with information on forecasts for traffic, accidents, and road closures, as well as recommendations for the quickest routes to their destinations. Siemens Mobility, for example, has developed a prototype AI monitoring system that uses traffic cameras to modify traffic signals depending on actual road jam, reducing traffic issues. Al will be able to decrease duplicate traffic, increase road safety, and shorten wait times in this manner.

FIGURE 9.4 Traffic pattern.

9.3.5 DETECTION OF AN ACCIDENT

Several people have attempted to determine the schedule, place, and intensity of an event to aid traffic management in reducing traffic congestion. In these efforts, manual reporting, automated algorithms, and neural networks were all employed. Human-written manual reports might cause a delay in discovering issues and can be costly [7]. Algorithms, on the other hand, can use information obtained from sensors along the path to estimate flow parameters before and after an incident. Because it is difficult to utilize an algorithm on major highways due to street road and traffic signals, neural networks techniques were developed. To determine the incidence of an event on a motorway, a neural network method was used to classify it. They discover an excellent software package for real-time detection of all vehicle objects.

Statistical techniques such as the California Algorithm have been used to construct detection algorithms, however utilizing an algorithm on arterial roads is difficult owing to street parking and traffic signals, therefore neural networks approaches have been developed. To determine the occurrence of an event on a motorway, a classification of neural network method was used. They discover a good software program for detecting all vehicle objects in real-time.

Using both virtual and real data, a more specific method called "efficient multiple model particle filter (EMMPF)" was developed to recognize traffic occurrences. Facebook and Twitter can be used to broadcast live events. It means that using Facebook to report traffic accidents on highways and arterial roads is an expense and encourage innovation.

9.4 MODELS OF PREDICTION

With the increasing expansion of ITS, new methods for forecasting traffic data are becoming more important. These technologies are essential for the success of sophisticated passenger data systems, modern road environment, sophisticated transport networks, and heavy truck operations. Historical data obtained from sensors deployed on highways was used to build intelligent prediction systems. Deep learning and Ai technologies use this information to build real-time, short-term, and long-term forecasts. With 90–94% accuracy, the algorithm was able to forecast speed for the next five minutes [8]. The inclusion of a contextual level distinguishes it from traditional feed forward networks. This context layer works as a memory box by storing prior data. Ireland's road traffic control provided the traffic volume data. The future traffic flow's outcome. This system is designed as a feed-forward deep network utilizing a bottom technique to detect the problem.

When the amount of hidden nodes is double that of input nodes, the network performs better. With a growth rate of 0.5 and a reducing number of iterations, flow accuracy should be between 92% and 98%. The ATOS

Scientific Community also created a pattern-based strategy (PBS). Learning algorithm is the first type of pattern recognition, in which branded input is combined with a specific output. Second, unstructured learning, which uses unlabeled data to find similarities and pick the appropriate output; and semi-supervised learning, which employs a small amount of dynamic panel data mixed with a big amount of unlabeled data to do pattern recognition analysis.

The city of Germany submitted data on the rate of flow and velocity and showed that speed forecasting accuracy was 93% and flow rate forecasting accuracy was 86%. As indicated in Figure 9.1, there are three phases to building a good prediction model. Six of the 24 convolutional and recurrent neural networks (RNNs) that were tested were helpful in identifying various spatial transport flow characteristics [9]. They constructed a basic RNNs for brief predictions using Jordan's neural network. The context layer distinguishes it from a traditional feed-forward network. This context layer functions as a memory box, holding previously acquired data. The information saved at time (t) is subsequently feed-forward to the hidden nodes, together with the input at period (t) (t–1). As a result, it's sometimes referred to as "Jordan's Sequential Network" since it helps the network analyze the subsequence.

The traffic volume data was obtained from Ireland's road traffic management system. The end outcome is the traffic flow in the future. ATOS Scientific Community also created a PBS. Moderately finding, in which only a small amount of supervised data is paired with a significant amount of unlabeled pattern recognition analysis, and unsupervised learning, in which unlabeled data is used to forecast a structure and select the right output. The city of Berlin submitted data on velocity and rate of flow, and the findings showed that speed forecasting accuracy was 93% and flow rate prediction reliability was 86%.

To create a strong prediction, it is necessary to recognize 3 stages, as shown in Figure 9.5.

All data sources must be analyzed and exploited in Figure 9.5 in order to integrate the advanced model into the industry. Many models can be used in the concept design phase to provide a more thorough evaluation of the performance by detecting failure sources and the time needed to complete the project's whole life cycle. Because the model in this phase must be constantly changed and enlarged to deliver the best results, the last phase of Figure 9.1 displays real-time asset performance prediction information.

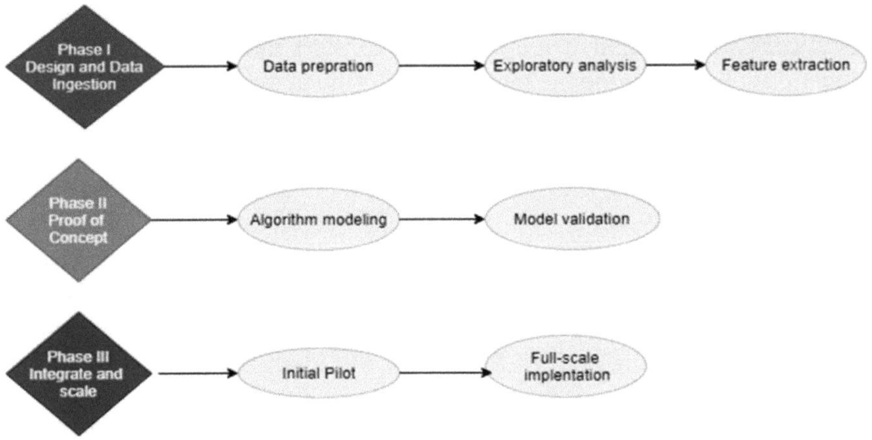

FIGURE 9.5 Stages of traffic management.

Data was initially grouped based on the daily route choices of the population. To categorize these data, a decision tree is employed, and a support vector machine (SVM) is used to enhance and extract more significant characteristics from the data. They're both supervised neural networks. To construct a forest, a decision tree employs a huge number of de-correlated decision trees. The greater the number of trees, the higher the accuracy. It operates by categorizing each tree (x) according to its credit value. The forest picks the category with the most votes, and the tree votes for it. SVM, on the other hand, classified the inputs by increasing the distance between them. They obtained data on large-scale ridesharing demand from DiDiChuxing in Guangzhou, China.

This result implies that the suggested model performs better. Using an inter feed-forward neural network, we projected taxi demand in Paris, China. They get their information through the Taxi Probe system, which equips cabs with data-gathering equipment (e.g., Location of the cab). Using 4 hours of historical demand data and 50 layers in the hidden nodes enhanced accuracy rate, according to the findings. Taxi drivers' performance L1-Norm SVM was used to identify the most essential feature from a cab pattern. Using taxi datasets from New York City, we employed ML algorithms to estimate taxi trip demand. The DNN outperforms other ML algorithms. To get ideal outcomes, the correct architecture must be identified. Finally, AI has the potential to save lives and decrease the effect of urban road accidents. The causes of automobile accidents differ in space and time. This study's findings demonstrate that this approach is successful and that it may be used to alert individuals about unsafe locations. To determine the risk level of traffic accidents, a Stack DE noise Auto encoder Simulation model was developed.

They discovered that by employing ML techniques, they may identify key hidden designs in previous car accidents. In Arab Emirates sustainable development 2019, 8 out of 24, it was discovered that gender (males have the highest accident rate), age (those between the ages of 18 and 30 are the most likely to be involved in an accident), and collision type and location all were linked to the fatal severity of an accident. Deep learning techniques such as tree structure and MLP were used to identify these factors.

9.5 AVIATION AND PUBLIC TRANSPORTATION AI APPLICATIONS

Field data collecting has become more dependable and approachable because of advances in computer science. Because public transportation data is voluminous and unstructured, it must be correctly modeled in order to fully exploit it. Aviation AI has been shown to improve the efficiency of aircraft journeys. Technology (ML), software/hardware, and applications may all benefit from AI (Intelligent Maintenance, Flight Route Optimization). When an airplane is landing, the unsupervised ML method is trustworthy for increasing safety [10]. In a nutshell, the plane's safety was analyzed by utilizing the Probabilistic neural network to verify the engine on board (PNN). The network training of the PNN is identical to that of the MLP, with the exception that it uses a back propagation algorithm instead of a linear function, and the RBN's hidden layer requires more neurons.

To better identify airplane turbulence, they develop an automatic supervised Random Forest approach. This technology can assist the pilot in staying on course, saving money on fuel, and improving air control management. It's critical to use ANN to improve the gas turbine monitoring system during flights. The sharing economy in mobility has proven desirable in congested urban centers due to its capacity to provide solutions to eliminate single or less occupant personal autos from the road, especially as it contributes to a reduction in inner-city traffic problems.

By adopting new adoptions to lessen traffic congestion and environmental effects, this enables for a win-win approach that gives both quick economic advantages and efficient long-term sustainable solutions. According to the definition of the collaborative consumption, "ICT-enabled networks for transfers of products and services drawing on non-market logics such as hosting, renting, donating, and exchanging as well as economic logic; renting and sale." Uber and Airbnb are the market's pioneers in the sharing economy, which is gaining popularity largely in the transportation and accommodation industries.

It's becoming more of a shared economy, with businesses attempting to develop business models for new shared mobility services in order to create a better mass transit system and more effectively bridge the demand-supply gap. Majority of customers are particularly interested in on-demand bike, car, and transit sharing. In addition, these shared mobility solutions address a major weakness in "first-mile, last-mile" transit.

9.5.1 URBAN MOBILITY WITH INTELLIGENCE

Smarter decision-making based on real-time data and network optimization through effective infrastructure utilization are among the goals of intelligent urban transportation in the future. Importantly, advocating for a better, safer, and healthier transportation system that promotes intelligent connectivity in order to build a long-term, seamless, and environmentally friendly network [11]. Autonomous vehicles (AVs), or automobiles that can travel without the help or supervision of a human driver, were on the horizon.

9.5.2 LAW ENFORCEMENT

Sensors and cameras placed along the roads collect a massive amount of data on traffic trends. This data is analyzed using big data and AI technology to discover traffic patterns. Thanks to the beneficial insights, commuters can get information on traffic forecasts, accidents, and road closures, as well as suggestions for the quickest paths to their locations. Siemens Transportation, for example, has developed a concept AI surveillance system based on the traffic sensors that alters traffic lights in real time to alleviate traffic congestion. AI will be able to decrease redundant traffic, increase traffic safety, and shorten wait times in this manner.

9.5.3 PREDICTIONS FOR DELAY

The delays in transportation, especially air travel, are another challenge. Figure 9.6 depicts flight delays, which in the United States might cost up to $39 million [12]. According to a research issued by the University of California, Berkeley, the United States is the most prosperous country in the world. AI turns to the air transport for help in order to avoid the costs of flight delays as well as the bad experience of passengers. AI may utilize its abilities to predict anything from indecency to a precise number of technical failures that could cause flight delays, using analysis of data and machine vision to assist and reduce passenger long waits. While AI and ML use and examine real aircraft data, records, and meteorological data, computer vision systems can watch planes on a periodic basis. Computing can expose certain hidden trends, allowing the sector to get important insights on potential delays or cancellations.

FIGURE 9.6 Delay protection.

9.5.4 TAXIS BY DRONE

Drones have already been utilized in delivery methods, but they could soon be included in vehicles as well. Unmanned aerial vehicles (UAVs) are a unique technology. Strategy for lowering carbon dioxide emissions, road congestion, and infrastructure costs [13]. Customers will benefit from drone taxis since they will arrive at their destinations faster and journey time will be reduced to a minimum. When it comes to topics like spatial regeneration and structural infrastructure development in metropolitan areas, drone taxis are frequently a reality. A demonstration of a driverless air vehicle in China, where 17 passengers got to enjoy efficient air mobility for the first time, was one of the most recent examples of drone taxis [14, 15]. Uber made history when it announced a collaboration with NASA to build the technology needed to operate autonomous "flying taxis" (Figure 9.7) a few years ago. Applications based on these use cases are likely to be similar in the future.

FIGURE 9.7 Design of drone.

9.6 THE LIMITATION OF ARTIFICIAL INTELLIGENCE (AI) TECHNIQUES

AI approaches have been subjected to a number of critiques. An overview of the transportation sector was presented to them. One of AI's major flaws is treating ANNs like a "black box." Without any prior understanding of the system's fundamental mathematics, the link between input and output is formed [16]. The ability of ANNs to generalize when some information is missing from data sets has been called into doubt. They overcame this restriction by using a hybrid method that combines neural networks with other traditional techniques and AI technologies to solve the problem. This need for hybridization to enhance performance, particularly in multi-scenarios, is seen as a universal flaw.

The development of AI-based for an effective transportation system is exceedingly tough due to the increase of mechanical intelligence as well as the proper interpretation of human-based information. AI applications in the transportation industry are limited to specific ITS functions like data processing and traffic predictions. It would be more efficient to have AI programs that could manage the entire operation. The accuracy and timeliness of forecasts degrade when AI techniques rely on data obtained using traditional methods such as sensing devices, detectors, and controllers.

The ability to employ AI to discover the optimum solution for Bitmap algorithms (e.g., GA and ACO). Unlike AI raster algorithms, techniques can give a genuine knowledge of the fundamental structure of the issue and the nature of the answer in addition to mathematical computations. In difficult optimization situations where standard mathematical approaches are impractical, the rapid analytical answers provided by these algorithms are preferable to no solutions. Building smart technologies is expensive due to their complexity and ongoing maintenance, and AI-driven systems lack user privacy and transparency, to name a few instances.

We'll look at a few AI computation complexity constraints in the next section. In a basic algorithmic scenario, the number of data increases at a constant pace as the number of operations increases. In complex algorithm setups, the amount of operations and calculations is expressed in time complexity (X2), logarithmic time (2X), and exponential time (X!). Algorithm complexity is expressed as a P problem or an NP problem. To deal with the complexity of the expanding data density, rapid processing equipment are required. To solve it, a non-deterministic turning machine is employed, and polynomial time methods are used to verify it.

The concept is to use a large number of computers to simultaneously propose all possible solutions to the same problem, with the best solution arriving in polynomial time. There are two types of NP: NP-hard and NP-complete. The NP-hard problem emerges when attempting to create an algorithm that works for all inputs, whereas the NP-complete problem is impossible to fix in constant time and hard to approximate in an optimization problem.

Traffic flow, speed, occupancy, and passenger behavior are examples of these properties. The initial data set is split into various subgroups and spread across numerous parallel processors. The computation will take longer to finish as a result of this. The map-reduce approach can be used on a single computer with multiple cores or CPUs. Each CPU or core will handle a portion of the massive datasets in order to generate the best reaction and speed up AI learning algorithms.

They demonstrated that it is a successful approach for adapting when identifying abnormalities in traffic occurrences for a large data set. This works by splitting the data into subgroups and analyzing each set separately for parallel processing. To find the bus arrival time, use map-reduce. To split the running duration of a bus, they utilize the Clustering K-Mean method.

The data may be clustered into k groups using the K-Mean method. It runs several times in order to identify multiple suitable configurations. Deep learning was used in picture identification jobs for the healthcare industry, as well as Facebook face recognition features, to arrive at this figure. ML [17] AI technology is helping the automotive, banking, and data mining industries better their operations. In addition to detecting patterns among data for future study, deep learning will see a significant increase in value, reaching $10.2 billion by 2025.

They talked on how AI's value in the travel and transportation industry may be expanded from a corporate, economic, and societal standpoint. AI's current deep learning neural networks, as demonstrated in Figure 9.8, can provide better value than previous techniques. Significant savings can be made in two important areas: travel and transportation-related services. AI techniques could be used to find the best and fastest route for the convenience of modes of transport and delivery businesses.

One European company was able to identify real-time truck performance and driver behavior by analyzing data from road sensors [20]. This resulted in a 15% reduction in gasoline costs as well as a 15-minute reduction in delivery time. By analyzing the weather and traffic congestion with AI algorithms, airlines can reduce the costs of cancellation.

Reduced traffic congestion, optimum public transit scheduling for consumers, self-driving car user safety, and enhanced air quality are among the additional advantages [18]. The value of the travel industry grows from 62% to 128% on average, while transport and logistics increases by 89%. Deep learning values will rise as AVs become more common. According to estimates, if 30% of vehicles in Australian cities are self-driving by 2030, the cost of congestion will drop from $38 billion to $26 billion. The cost of traffic congestion in Australia after deploying 30% AVs on the road as shown in Figure 9.9.

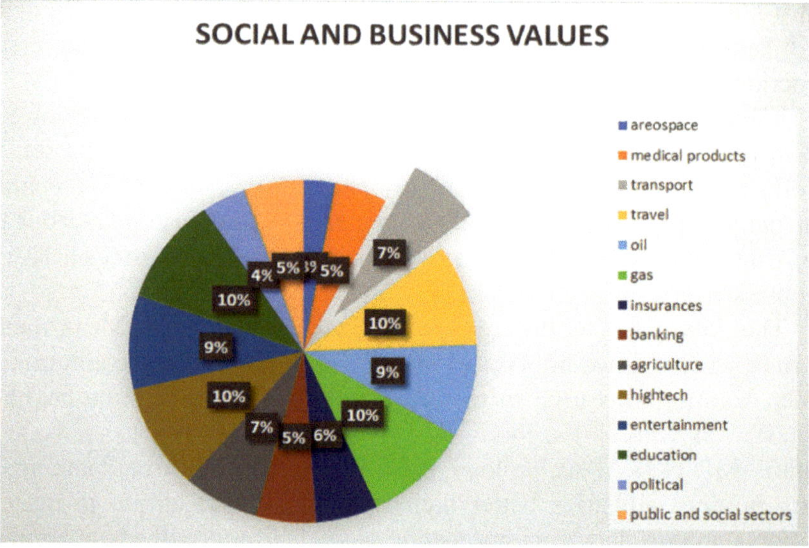

FIGURE 9.8 Benefits of AI-assisted performance.

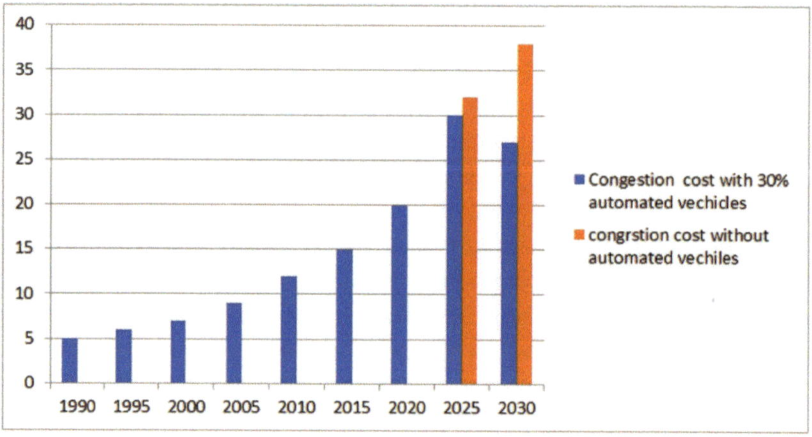

FIGURE 9.9 The cost of traffic congestion in Australia after deploying 30% autonomous vehicles on the road.

9.7 ADVANTAGES OF USING AI IN TRANSPORTATION

Despite the fact that the usage of AI varies, between locations, the best fit between AI and transportation comes naturally. Adoption of these technologies may have a huge influence on the whole sector [17]. Increased usage of AI will result in lower labor costs and better profitability, with fully automated fleets addressing the issue of long driving hours and rest stops. AI may additionally have a significant influence on road Accidents and safety. Driving in the dark might be a dangerous problem, but clever unmanned vehicles can help greatly. Autonomous cars that can run without human intervention can let drivers sleep while avoiding accidents. AI techniques allow us to estimate transportation utilizing travel information and data about current city activity, as well as automate the suggestion of other routes, making transport management more efficient. AI might benefit complex infrastructures and many elements of collaboration chains, such as an optimum travel schedule, reduced waiting time, real-time traffic detection for route changes, and so on. Data analytics could also aid in improving logistics transport systems and improved security. There are many more advantages to mention, but because AI is still developing, the advantages will continue to evolve (Figure 9.10).

FIGURE 9.10 Advantages of using AI.

9.8 DRAWBACKS OF AI IN TRANSPORT

Although AI has several advantages for every sector, there are also disadvantages to consider, such as increasing transparency in AI decision-making or safety concerns with autonomous cars. Job flow is one of the disadvantages, since it has become a big concern for truck or taxi drivers, as well as other industry participants [19]. Despite the efforts of social specialists to explain that employment abilities may be transferred to other industries, conflicts persist. All implementations might have downsides, as developing nations would have a difficult time implementing such solutions since their infrastructure will not be able to provide upkeep. Transportation firms also suffer potential AI disadvantages is shown in Figure 9.11, since transportation expenses account for up to 10% of a company's income. To remain competitive in their industry, all existing businesses will need to develop and implement AI technologies, and the enormous costs of manufacturing, fixing, and sustaining such sophisticated machinery can be too expensive. Companies introducing If AI to transportation [21] wants to win consumer trust, they must take AI's ethics seriously. AI technology has a long way to go before it approaches human intellect.

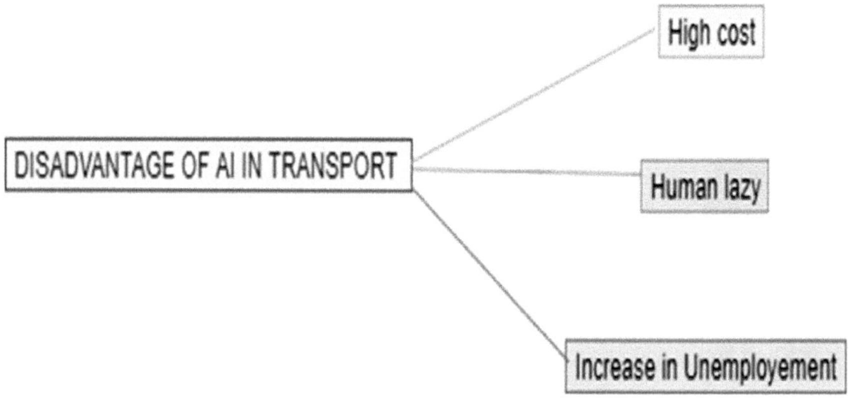

FIGURE 9.11 Disadvantage of ai in transport.

9.9 CONCLUSION

AI has progressed from a futuristic concept to a part of our everyday life that we utilize without even realizing it. Our mobile applications, social networking feeds, and the way Grammarly evaluates our grammatical errors all contain AI. For a while, the transportation sector has employed certain AI solutions, but it won't be long until AI is used more extensively in transportation and logistics. It'll only be a matter of time until we witness the bright future made possible by AI it becomes more subtle with time.

KEYWORDS

- **artificial intelligence**
- **intelligent transportation systems**
- **joint program office**
- **machine learning**
- **network design problem**
- **social networking**

REFERENCES

1. Weinan, Z., Haiming, J., Lingyu, Z., Hongtu, Z., Jessie, Z. L., & Jieping, Y., (2019). CIKM 2019 workshop on artificial intelligence in transportation (AI in transportation). In: *Proceedings of the 28th ACM International Conference on Information and Knowledge Management (CIKM '19)* (pp. 2995, 2996). Association for Computing Machinery, New York, NY, USA. doi: https://doi.org/10.1145/3357384.3358802.
2. Peter, K., Jinglu, S., Mingzhou, J., Helmut, S., & Christopher, C., (2019). Evaluation of operational and environmental sustainability tradeoffs in multimodal freight transportation planning. *International Journal of Production Economics, 209*, 411–420. ISSN 0925–5273, https://doi.org/10.1016/j.ijpe.2018.08.011.
3. Ivica, D., Tamara, P., Damjan, M., Zdenko, K., & Juraj, O., (2020). Highly scalable traffic management of autonomous industrial transportation systems. *Robotics and Computer-Integrated Manufacturing, 63,* 101915. ISSN 0736-5845, https://doi.org/10.1016/j.rcim.2019.101915.
4. Chavhan, S., Gupta, D., Chandana, B. N., Khanna, A., & Rodrigues, J. J. P. C., (2020). IoT-based context-aware intelligent public transport system in a metropolitan area. In: *IEEE Internet of Things Journal* (Vol. 7, No. 7, pp. 6023–6034). doi: 10.1109/JIOT.2019.2955102.
5. Aljoufie, M., & Tiwari, A., (2021). Citizen sensors for smart city planning and traffic management: Crowd sourcing geospatial data through smartphones in Jeddah, Saudi Arabia. *Geo Journal.* https://doi.org/10.1007/s10708-021-10423-4.
6. Zhu, F., Lv, Y., Chen, Y., Wang, X., Xiong, G., & Wang, F. Y., (2020). Parallel transportation systems: Toward IoT-enabled smart urban traffic control and management. In: *IEEE Transactions on Intelligent Transportation Systems* (Vol. 21, No. 10, pp. 4063–4071). doi: 10.1109/TITS.2019.2934991.
7. Alkinani, M. H., Khan, W. Z., & Arshad, Q., (2020). Detecting human driver inattentive and aggressive driving behavior using deep learning: Recent advances, requirements and open challenges. In: *IEEE Access* (Vol. 8, pp. 105008–105030). doi: 10.1109/ACCESS.2020.2999829.

8. Azzedine, B., Yanjie, T., & Peng, S., (2020). Artificial intelligence-based vehicular traffic flow prediction methods for supporting intelligent transportation systems. *Computer Networks, 182*, 107484. ISSN 1389-1286, https://doi.org/ 10.1016/j.comnet. 2020.
9. Ishwarya, S., & Madasamy, A., (2020). Deterioration models for prediction of remaining useful life of timber and concrete bridges: A review. *Journal of Traffic and Transportation Engineering (English Edition), 7*(2), 152–173. ISSN 2095-7564, https://doi.org/10.1016/j.jtte.2019.09.005.
10. Hubbard, S. M. L., & Hubbard, B., (2019). A review of sustainability metrics for the construction and operation of airport and roadway infrastructure. *Front. Eng. Manag., 6*, 433–452. https://doi.org/10.1007/s42524-019-0052-1.
11. Kircher, R., Kluehspies, J., Fritz, E., & Koehler, R., (2019). Mobility on demand – vision of a flexible future urban mobility. *Transportation Systems and Technology, 5*(2), 130–138. doi: 10.17816/transsyst201952130-138.
12. Ping, H., Chao, W., Liping, F., Javad, L., Chaozhe, J., Qiyuan, P., & Xinyue, X., (2020). Modeling train operation as sequences: A study of delay prediction with operation and weather data. *Transportation Research Part E: Logistics and Transportation Review, 141,* 102022, ISSN 1366-5545. https://doi.org/10.1016/j.tre.2020.102022.
13. Robin, K., Tobias, B., & Liliann, F., (2020). Drones for parcel and passenger transportation: A literature review. *Transportation Research Interdisciplinary Perspectives, 4*, 100088. ISSN2590-1982. https://doi.org/10.1016/j.trip.2019.100088.
14. Vanitha, C. N., Sridhar, K. L., & Dhivakar, R., (2021). Automation of noise detection using the internet of things. *Proceedings of the 6th International Conference on Inventive Computation Technologies: ICICT 2021* (pp. 184–189). this link is disabled, 9358628.
15. Mehta, P. L., Kalra, R., & Prasad, R., (2021). A backdrop case study of AI-drones in Indian demographic characteristics emphasizing the role of AI in global cities digitalization. *Wireless Pers. Commun., 118*, 301–321. https://doi.org/10.1007/s11277-020-08014-6.
16. Rajesh, K. D., Lalitha, K., Anitha, S., Khaitan, S., Gupta, P., & Goyal, M. K., (2021). Hybrid and dynamic clustering-based data aggregation and routing for wireless sensor networks [JB]. *Journal of Intelligent & Fuzzy Systems*, 1–15.
17. Vanitha, C. N., Jeevaa, V. N., & Shriman, S. P., (2019). Image and face recognition using CV lens machine learning android application. In: *2019 5th International Conference on Advanced Computing and Communication Systems: ICACCS 2019* (pp. 972–975).
18. Aalok, K., & Ramesh, A., (2020). Assessment of environmental and social sustainability performance of the freight transportation industry: An index-based approach. *Transport Policy*. ISSN 0967-070X. https://doi.org/10.1016/j.tranpol.2020.01.006.
19. Saravanakumar, P., Sundararajan, T. V. P., Rajesh, K. D., Yunyoung, N., & Seifedine, K., (2021). Ruzicka indexed regressive homomorphic ephemeral key benaloh cryptography for secure data aggregation in WSN. *Journal of Internet Technology, 22*(6), 1287–1297.
20. Vanitha, C. N., Sriyazhini, L., Sathya, D., & Devayani, S., (2021). A methodical procedure for home autonetics system using internet of things with multi-functional sensors. *Lecture Notes in Electrical Engineering, 700*, 261–269.
21. Igor, H. T., Stéphane, G., Vivient, C. K., Jean, C. K., Yazan, M., Amro, N., & Vincent, H., (2020). A critical review of the use of holonic paradigm in traffic and transportation systems. *Engineering Applications of Artificial Intelligence, 90*, 103503. ISSN 0952-1976, https://doi.org/10.1016/j.engappai.2020.103503.

CHAPTER 10

AI Techniques for Future Smart Transportation

S. ANBUKKARASI and C. R. DHIVYAA

Assistant Professor, Department of Computer Science and Engineering, Kongu Engineering College, Tamil Nadu, India

ABSTRACT

Smart transportation system (STS) is an emerging area of research, and it draws the attention of academicians and Industrialists. It contributes various services to end users, including road safety and accident prediction, traffic prediction and control, and vehicle control and monitoring. The environmental pollution, heavy traffic, unpredictable congestion increases the complexity of the STS. Henceforth, the future STS is in the position to provide solutions for all these problems. It paves the way to enhance congestion control, CO_2 emission control, user security, etc., as part of the transportation services. The increase in population and the transportation needs of the individual makes the Intelligent Transport System embrace various techniques of Artificial Intelligence (AI). In this chapter, we mainly focus on the various artificial techniques for future smart transport management systems. In the first section, the basic existing transportation system and their problems are discussed. The second section discusses the different kinds of AI technologies which deal with smart transportation. The third section discussed the emerging AI technologies for the future smart transportation services. The last section discusses some of the new AI technologies which handle the future smart transportation services.

Artificial Intelligence for Future Intelligent Transportation: Smarter and Greener Infrastructure Design. Rajesh Kumar Dhanaraj, Nilayam Kumar Kamila, Subhendu Kumar Pani, Balamurugan Balusamy, & Vani Rajasekar (Eds.)

10.1 EXISTING TRANSPORTATION SYSTEM

The process of transporting goods, people or something else from one place to another place is known as transportation. The increase in population makes the transportation problem a tedious one. All over the world, People use motor vehicles in billions for transportation purpose in day-to-day use. The vast usage of these vehicles creates so many problems in managing transportation efficiently. In this section, we discuss the existing transportation system and the problems we face due to this system. The transportation system is considered to be smart when it is supposed to do all or any one of the following tasks:

- Pollution control;
- Traffic management system;
- Driver assistance system;
- Navigation and vehicle information system;
- Accident control.

For addressing these issues, smart transportation system (STS) should involve many techniques such as real time data creation, communication, data mining and AI, ML and database management. By using all these technologies, STS provides efficient solutions for the problems discussed above.

The application of STS can be given in broad category as given below:

- Accident prediction and road safety;
- Vehicle control;

The function of STS is divided into sub tasks as given in Figure 10.1. STS deals with real time data which may be collected from various sensors.

10.1.1 TRAFFIC PREDICTION AND CONTROL

Many works have been carried out so far in smart transportation which deals with one or more problems of transportation. Genetic algorithm (GA) is used for controlling the movement of traffic by controlling the flow route. This algorithm helps to adjust the timings of the signal lights. With the simulation result, they showed that air pollution assessment can be automated [20]. Traffic control is achieved by GA by predicting the time at which vehicles are usually stopped at intersections [8]. Vehicle-to-infrastructure (V2I) communication is helpful in identifying the time for vehicles to arrive at the intersections. The data is collected from the vehicles themselves and GA helps to determine the time to be delayed for arriving at the intersection. By this means, the system updates each vehicle, and the vehicle can make the plan for the speed profile.

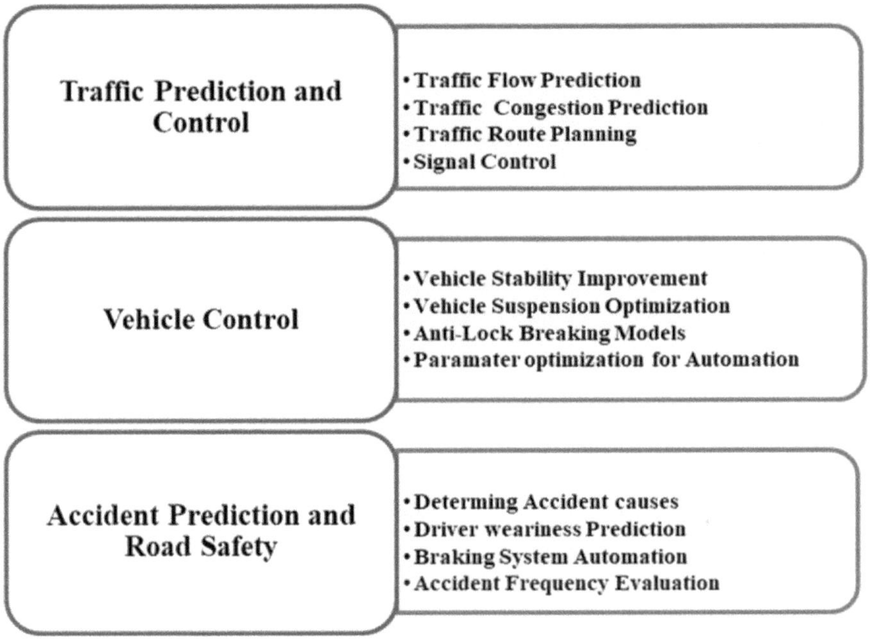

FIGURE 10.1 Tasks of STS.

An intelligent traffic control system is proposed in Ref. [11] with radio frequency identification (RFID). RFID tag is placed in the vehicles which cannot be easily removable, and it can be taken by RFID readers. Number of vehicles at particular time in the traffic can be counted by RFID reader which gives the network congestion in a given time. Based on this information, the duration of green light can be decided. The authors claimed that usage of RFID tag gives the best results in traffic control management. The basic architecture of smart traffic monitoring methodology is depicted in Figure 10.2 as discussed in Ref. [4].

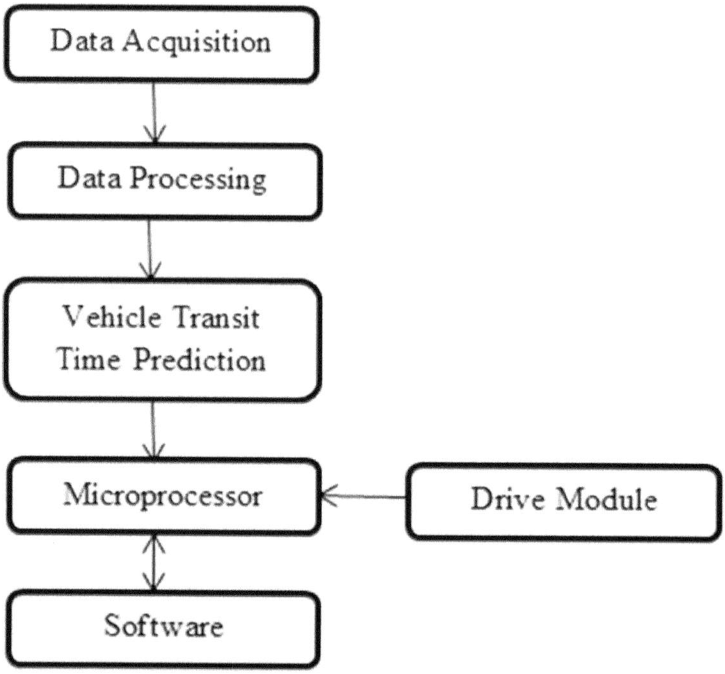

FIGURE 10.2 Basic design model of smart traffic prediction system.

The other technique known as expert system (ES) is used to control the traffic [12]. The modified algorithm known as belief rule based expert system (B-RBES) is used to handle the traffic signal based on the uncertainties. The hand-crafted defined rules are used to make the decision in controlling the traffic signal. The major drawbacks of these kinds of rule-based system are the extensive list of rules to be accompanied by the system for the predictions.

10.1.2 VEHICLE CONTROL

Vehicle control system (VCS) obtains the input parameters from the outside environmental variables, vehicle positioning and sensor information. It processes all these input variables and control brake, steering, and acceleration of the vehicle. VCS mainly focuses on the comfort level of Passenger and driver, vehicle safety, efficient usage of fuel, etc. The general structure of VCS is given in Figure 10.3. GA is proposed for charging the electrical vehicles [13]. This work mainly focuses on the efficiency of the charging control system. The system is designed by the considering the power constraints. For vehicle monitoring system, a system is prosed to provide exact GPS information [18]. The quality of GPS positioning is monitored by integrity check at each level. Map matching process is improved by checking the integrity of velocity using Doppler information of GPS. Fuzzy logic is used for controlling the vehicle along with the human Intelligence for controlling the vehicle's direction and speed [22] In their proposed work, they used global positioning system (GPS) and artificial vision with wireless local area network (WLAN).

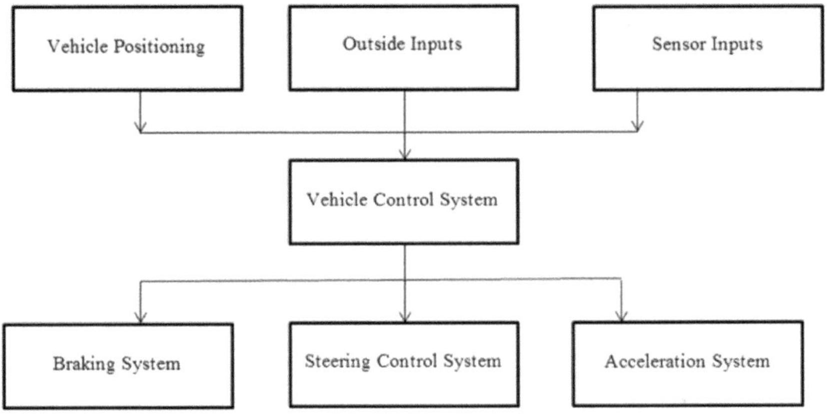

FIGURE 10.3 Vehicle control system.

ES is applied in vehicle driving simulator (VDS) in which users can be able to feel the driving by upgrading the database known as knowledge base (KB).

10.1.3 ACCIDENT PREDICTION AND ROAD SAFETY

Accident prediction and road safety system of STS covers the following criteria:

- Prevention of road accidents;
- Analyzing the causes for road accidents;
- Alleviate the reason for accidents so that it improves the survival of the drivers and passengers in the vehicle.

Accidents are caused by various reasons which involves careless of drivers, road condition, negligent driving of other drivers, etc. To avoid such crucial accidents, it is necessary to gather the input values from the in-vehicle sensors and process those values to alleviate an accident, predict the accident, etc.

Using the external factors of the accident, the severity of the injury for the driver and passengers are predicted in Ref. [24]. One of the ML techniques known as Artificial Neural Network is used for predicting the severity. Disaster management is improved by incorporating intelligent decision support system in Ref. [14]. The system gathers the live data from various sensor networks. It integrates various data such as traffic, road network, and meteorological data for giving intelligent transport platform by implementing intelligent models and algorithms. When the accident is occurred, existing resources are allocated in optimal way using multi-objective GAs in Ref. [17]. The major aim of the work is to reduce the severity of the injuries.

The general workflow of STS for accident prediction and road safety is given in Figure 10.4.

An ES based on web is developed to reduce the collision between the pedestrian and vehicles. Fuzzy logic system in terms of toolbox is embedded in MATLAB by investing many parameters including residential volume, traffic density, Speed of the vehicle. Web based graphical user interface (GUI) is developed to evaluate the system. Many algorithms such as ES, fuzzy system are used to identify the fatigue level of the driver to predict the accident. Thus the system alerts the driver and avoids the accident.

10.2 VARIOUS AI TECHNOLOGIES FOR SMART TRANSPORTATION SYSTEM (STS)

One of the major reasons of immediate death around the earth is road accidents. Traffic accidents are a major threat to public health issues and also it causes economic damage to our nation. The reasons for road accidents are raising the population, growing number of vehicles all over the region and it increases the number of wounds in traffic accidents. Hence, the advancement of technology is crucial for improving the road safety by avoiding the possible road accidents. One of the emerging technologies is AI which has been applied for improving the performance of various applications of industries and business in real time. In recent days, AI techniques are needed to solve the transportation problem such as improving travel demand, environmental degradation, CO_2 emissions, safety measurements and the AI techniques are implemented to enhance the intelligent transportation systems (ITS). There are various Artificial Techniques available to address the issues and problems of transportation like safety management, traffic management, mobility management, accident prediction and public transportation management. The several AI techniques are applied in intelligent transport systems, and it is shown in Figure 10.5.

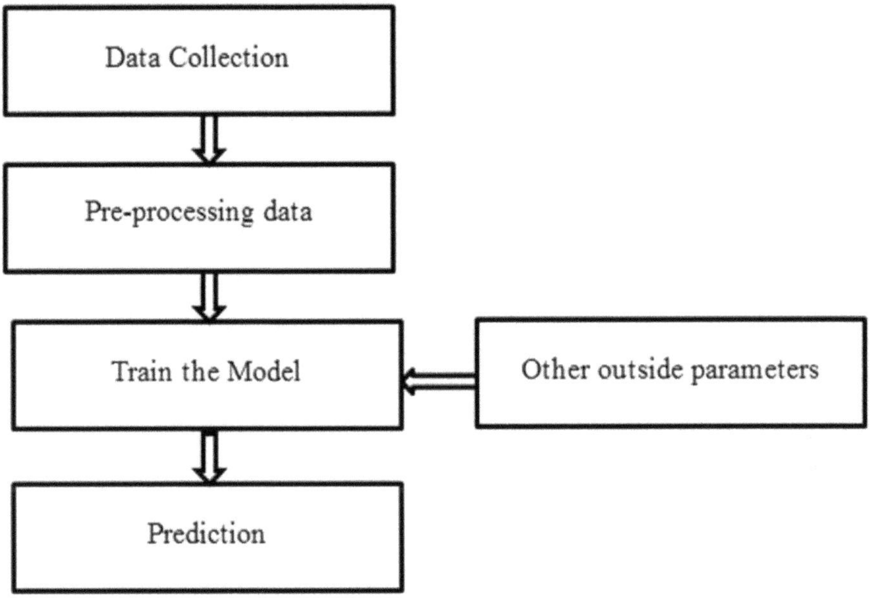

FIGURE 10.4 Transportation system for accident prediction and road safety.

10.2.1 ARTIFICIAL NEURAL NETWORK

The network contains neuron layers which are classified into three layers namely input, output, and hidden layers. Input layer gets input data from the user, the hidden layer acts as a middle layer between input and output layers and this layer performs all mathematical computations, at last output layer provides appropriate outcome for the given user's input data. In an intelligent transport system, the ANN model is applied for classifying the congestion level, predicting the traffic flow in urban areas, detecting accidents, detecting the location of vehicles, estimating the speed of vehicles on the road and so on. It provides a better solution for planning, scheduling, and developing the traffic signals for public ITS by analyzing the congestion level of the traffic on the road.

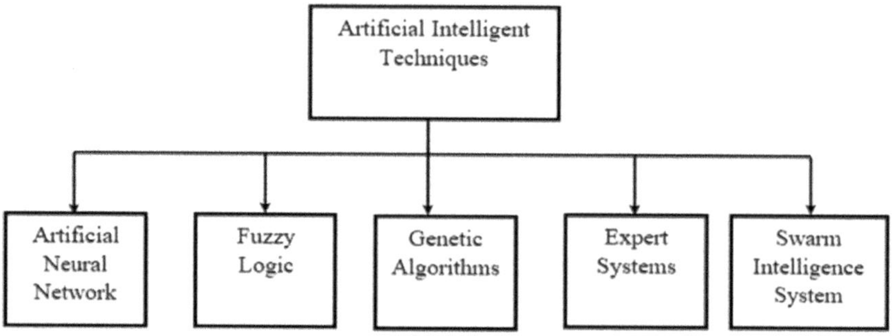

FIGURE 10.5 AI-based techniques used in intelligent transport systems.

The ANN model consists of data collection, transformation, and information processing units. In the data collection unit, all the traffic data are collected using sensors by capturing the number of vehicles and speed of vehicles on the road. The collected data are aggregated and sent to the remote data analysis. The average speed of the vehicles and traffic density of the road are calculated using the traffic raw data which are collected from the different fields. Finally, the information processing unit sends the processed traffic data into ANN model for predicting the congestion level of the traffic on the road. Figure 10.6 depicts the basic artificial neural network model for intelligent transportation system.

10.2.2 FUZZY LOGIC

Fuzzy logic is one of the AI methods where it takes decision as like human's decision making and it provides intermediate possible values between yes and no which is based on "If-Then" rules. In an intelligent transportation system, the fuzzy logic [19] is suited to address the problems of transportation like traffic prediction, traffic control, finding alternate routes and capturing the driver behavior. The basic model of fuzzy logic is shown in Figure 10.7. The fuzzy model consists of fuzzifier, intelligence (decision-making unit), rules and defuzzifier. The fuzzier performs the fuzzification process that transforms the crisp inputs of the system into fuzzy sets. The intelligence unit mimics the process of human reasoning based on the If-Then rules defined by experts. The fuzzy set attained by inference engine is transformed by defuzzifier into a crisp value. Some of the traditional existing controllers are not suitable for addressing the dynamic issues such as weather condition, road condition, and traffic jams. The fuzzy logic has the ability to enhance the existing traffic controllers to overcome the problem of dynamic conditions effectively. For instance, if speed is low and distance is low then break rate is low. The fuzzy logic takes the decision based on fuzzy If-Then rule as like human observation.

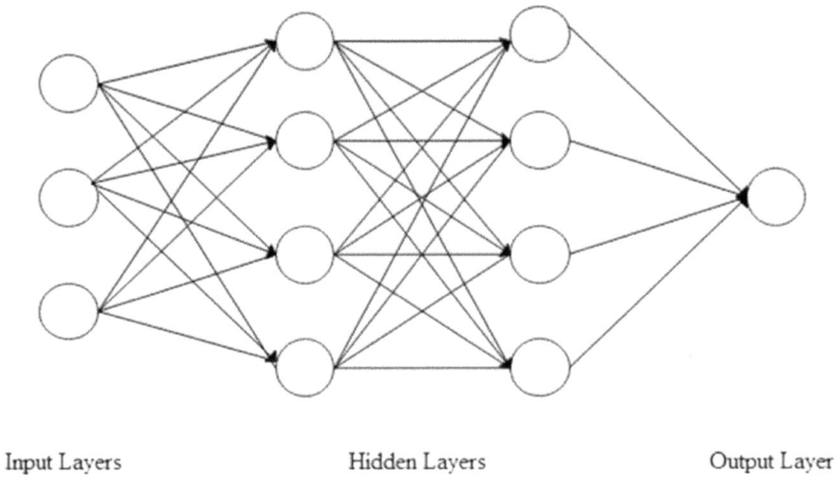

FIGURE 10.6 ANN model for intelligent transportation system.

10.2.3 GENETIC ALGORITHM (GA)

The popular artificial technique is GA [20, 21] which will be helpful for STS. The GA is efficient for guiding the driver to the target location by providing the information about the traffic on the road.

Figure 10.8 depicts the sequential approaches of the GA, and it is described as below:

- Preparing the initial value of the population. Each chromosome contains number of genes;

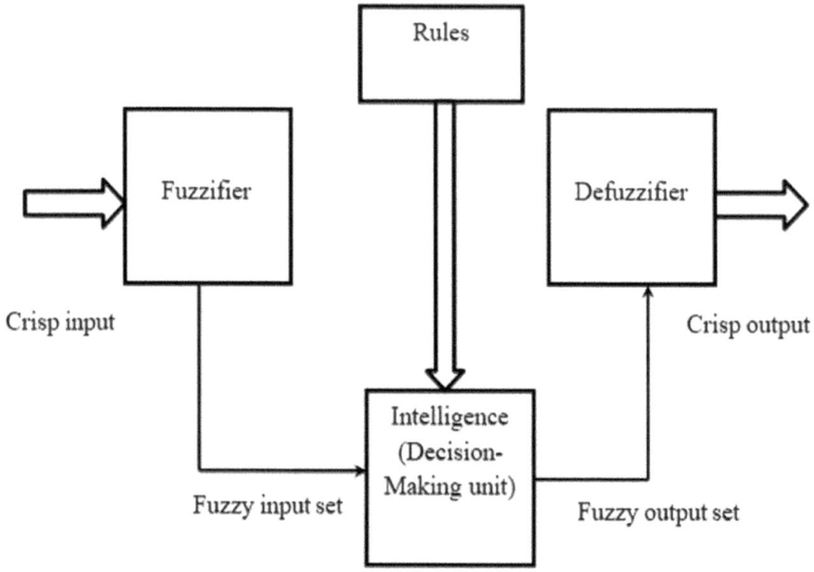

FIGURE 10.7 Fuzzy logic basic model.

- Evaluating fitness function for each chromosome in the population;
- Cross over operation is responsible for interchange the information between genes of the chromosome and it produces the new offspring;
- Mutation is applied to maintain the genetic divergence from one generation to another generation and it is also responsible for the new information;
- Calculate the fitness function for the new generation by computing the crossover and mutation process;
- Select the best chromosome among the other chromosomes and transfer to the next generation;
- Repeat the process until the condition is met.

The GA can be able to produce an approximate solution to address the problem of intelligent transport systems effectively.

10.2.4 EXPERT SYSTEMS (ESS)

In AI, the ES is a program which imitates the decision making capability of human experts based on predefined knowledge. It uses a knowledge procedure to solve the complex problems by utilizing human knowledge for analysis. It is a rule-based approach. The ES [6] is a crucial technique for vehicle maintenance, external vehicle's control in transit, vehicle scheduling and traffic signal control. The basic architecture of the ES is depicted in Figure 10.9.

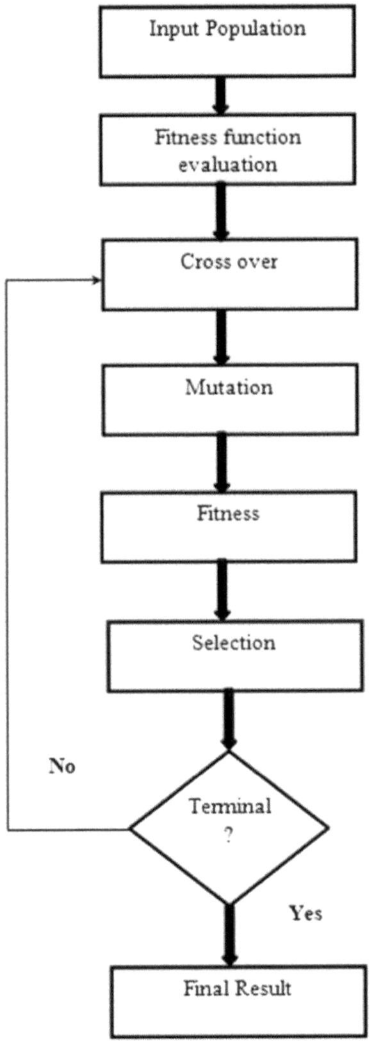

FIGURE 10.8 Sequential approach of genetic algorithm.

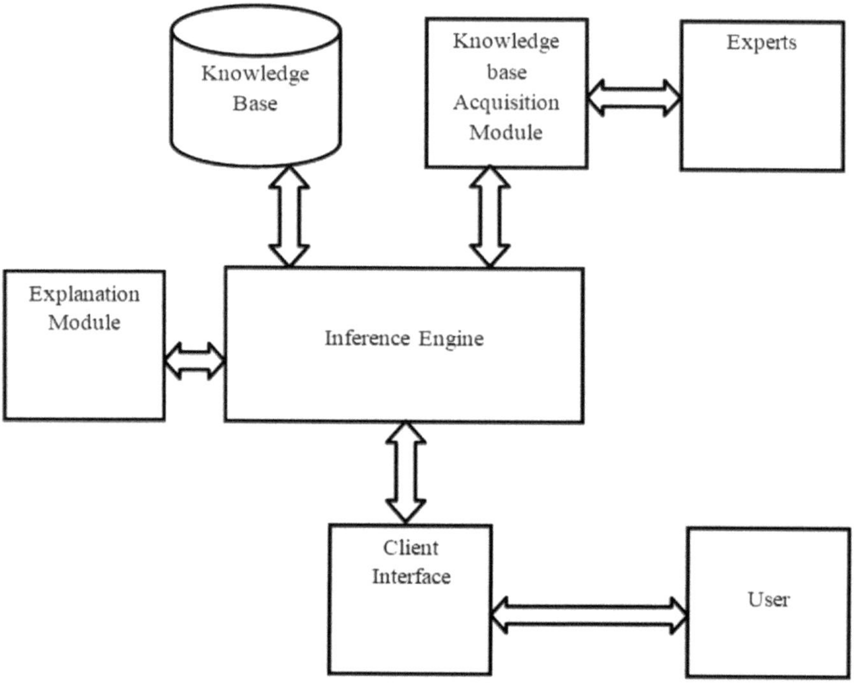

FIGURE 10.9 Basic model of expert system.

The ES consists of several components listed below:

1. **Inference Engine:** It performs actual reasoning process of knowledge, and it solves the problem by combining the information obtained by the KB and by the user. It is the heart of the ES.
2. **Knowledge Base (KB):** It is the repository of domain specific knowledge collected from the experts through acquisition module. It comprised of rules, facts, and other knowledge for providing the problem-solving solution by utilizing the AI concepts.
3. **Knowledge Base Acquisition Module:** It provides the knowledge to the knowledge database. The experts can enter and edit the set of rules (knowledge) in an ES. It gathers the intelligence information from the human experts and stores it into the KB.
4. **Explanation Module:** It explains the reasoning process, provides solutions and other information to the client. It also helps the experts to find the ability of the system's reasoning in order to reach a goal.

5. **Client Interface:** The user can access the system through the client interface. It acts as a mediator tool that provides a connection between the user and the system.

10.2.5 SWARM INTELLIGENCE SYSTEM

It is an AI system which consists of agents interacting locally with each other in their environment. Ant colony optimization (ACO) [10] and bee colony optimization [9] (BCO) are examples of swarm intelligent systems. ACO is an AI algorithm that contains a set of artificial ants to identify the optimal solution for a given problem based on the ant's foraging behavior. BCO is also an AI technique to detect good solutions for the problem based on the behavior of honeybees. ACO and BCO is a suitable model for traffic control systems to minimize the time delay, average waiting time and congestion level of traffic.

10.2.6 AI TECHNIQUES USED IN INTELLIGENT TRANSPORTATION SYSTEMS (ITS)

The different AI techniques have been used in Intelligent Transport Systems for controlling the accident and traffic, predicting the accident, road safety and traffic prediction.

Figures 10.10–10.12 depict the implementation of AI techniques in vehicle control, accident prediction and traffic control, respectively.

The various AI techniques have been implemented in various areas of the transportation system and it is listed in Table 10.1 and shown in Figure 10.13.

10.3 EMERGING AI TECHNOLOGIES FOR THE FUTURE SMART TRANSPORTATION SERVICES

10.3.1 DEEP LEARNING TECHNIQUES

Deep learning is a sub class of ML which deeply analyzes the huge amount of data generated by devices and provides a better solution for the complex problem. Hence, deep learning techniques are suitable for addressing the problems in ITS and it also helps for predicting performance of traffic,

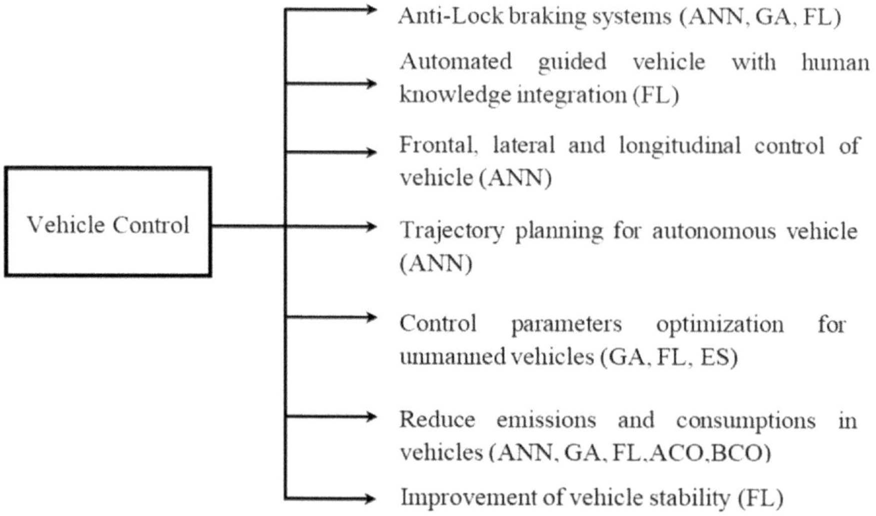

FIGURE 10.10 Artificial intelligence techniques for vehicle control.

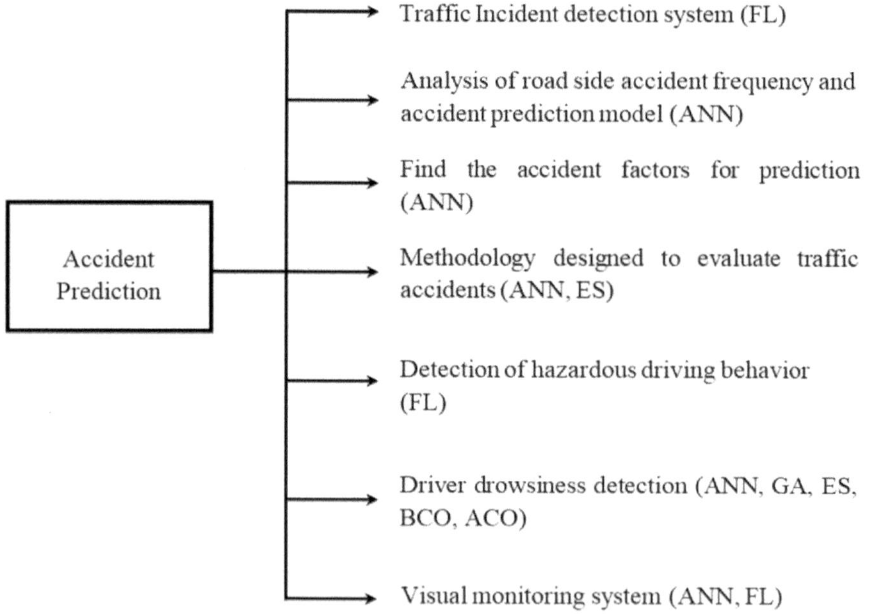

FIGURE 10.11 Artificial intelligence techniques for accident prediction.

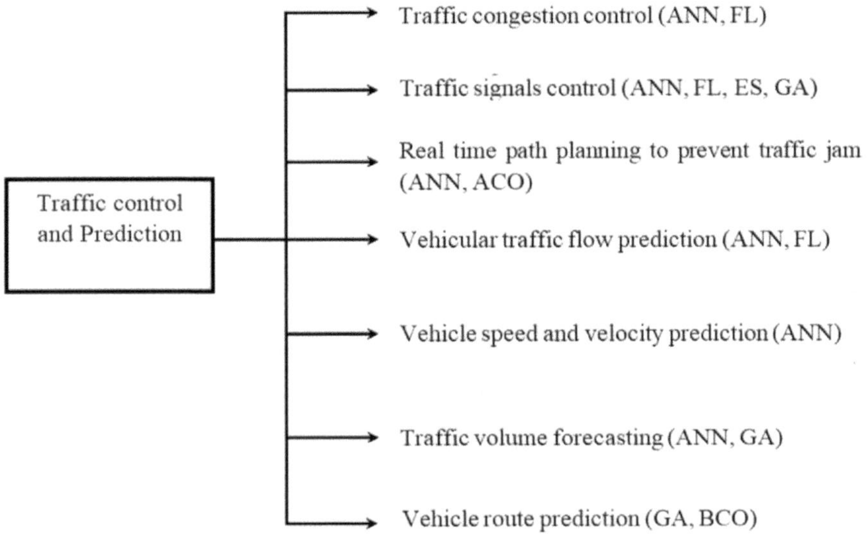

FIGURE 10.12 Artificial intelligence techniques for traffic control and prediction.

TABLE 10.1 AI Methods Used in Existing Transportation Systems

Different Areas of Transportation Systems	**Artificial Intelligent Techniques**				
	ANN (Artificial Neural Network)	**FL (Fuzzy Logic)**	**ES (Expert System)**	**GA (Genetic Algorithms)**	**Swarm Intelligence System (ACO, BCO)**
Vehicle control	4	5	1	3	2
Accident prediction	5	3	2	1	1
Traffic control and prediction	6	3	1	3	2
Total	**15**	**11**	**4**	**7**	**5**

detecting lane, identifying the best routes for transportation of tangible goods, preventing accidents, finding the objects near vehicles to improve the safety of transportation system, solving all the related issues of transportation system in an efficient manner.

The different deep learning techniques are listed below:

- Convolutional neural network;
- Recurrent neural network;

- Long short-term memory;
- Bi-directional long short-term memory;
- Gated recurrent unit (GRU);
- Attention based GRU.

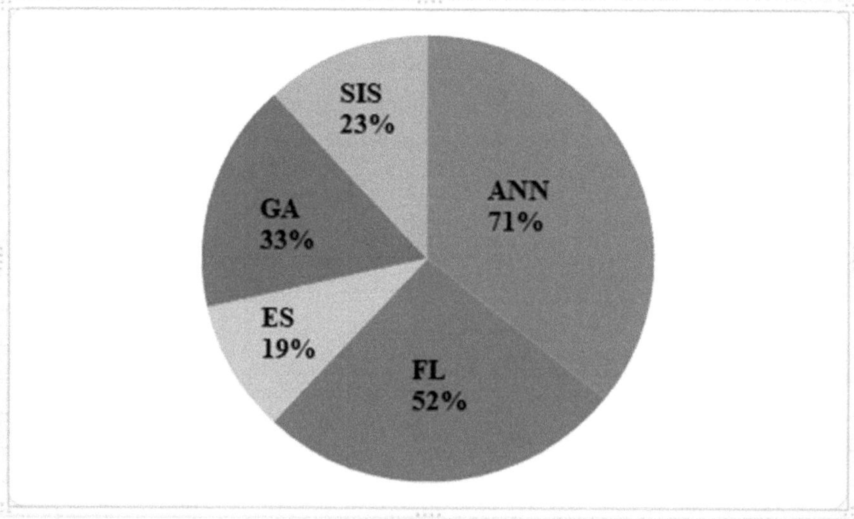

FIGURE 10.13 Comparison of various AI techniques in intelligent transport systems.

10.3.2 CONVOLUTIONAL NEURAL NETWORK (CNN)

The network [2, 3] contains 5 layers namely input, convolution, pooling, fully connected and output layers. The input layers are used to accept the inputs from the user, the convolution and pooling layers are used to detect the features from the given input, and finally, the fully connected and output layers are acted as a classifier to predict the class depends on the features. In intelligent transport systems, the convolutional neural network (CNN) is suitable for traffic sign identification, classification of vehicle, detecting congestion level of traffic and detecting pedestrian on the road. Figure 10.14 depicts the basic architecture of CNN.

In CNN, the activation function is one of the important factors where it decides which neuron will be activated and transferred to the upcoming layers. The various activation functions are shown in Figure 10.15.

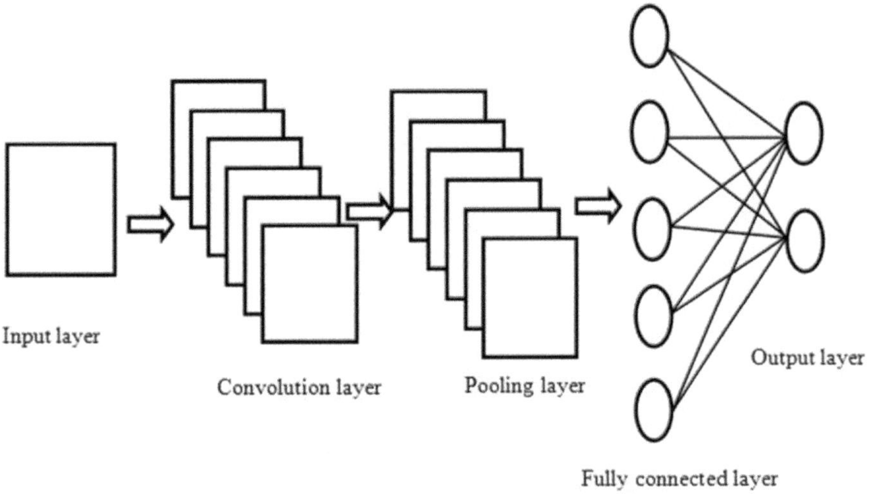

FIGURE 10.14 Architecture of CNN.

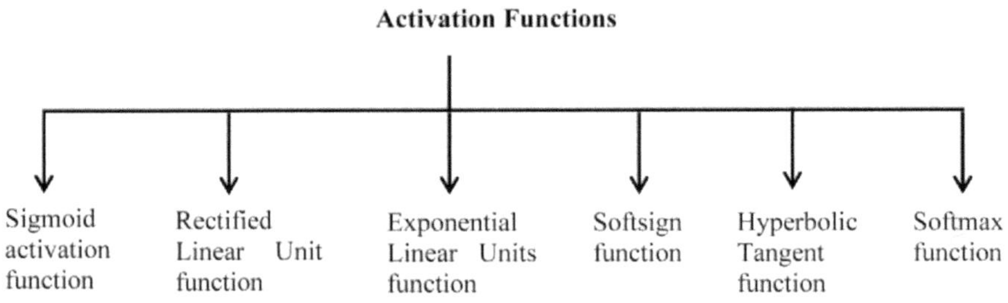

FIGURE 10.15 Activation functions.

There are several CNN models available to upgrade the system's performance. The best model will be selected based on the performance to predict the output class effectively. The different architectures of CNN are listed in Figure 10.16.

10.3.3 RECURRENT NEURAL NETWORK (RNN)

Recurrent neural network (RNN) is a kind of supervised deep learning technique which is used for capturing the sequences of data dynamically. In RNN [1], the output of the present input depends on the computation of the previous step; it copies the output and sends back to the RNN.

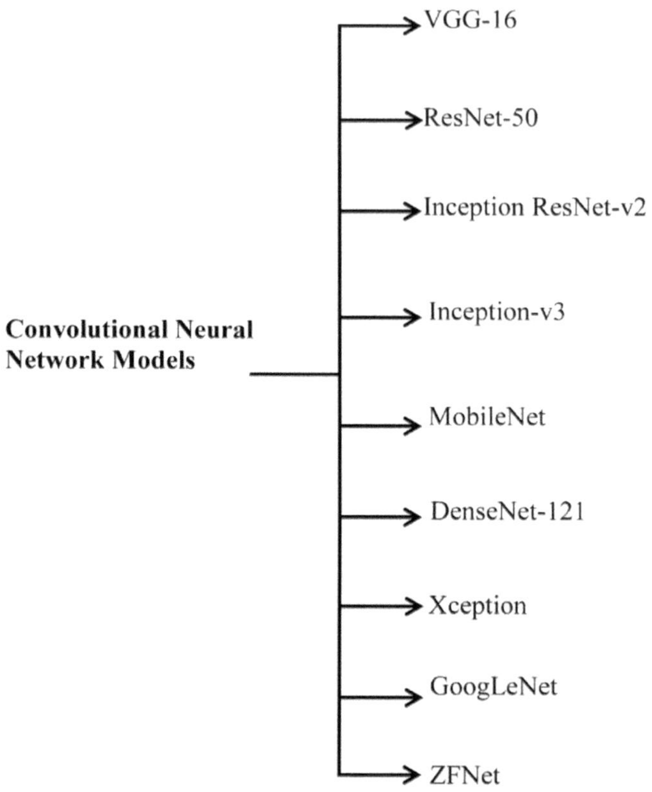

FIGURE 10.16 Different models of convolutional neural network.

The recurrent network has an internal memory to remember the past computations and it uses two inputs such as the current input and the recent previous. RNN contains a sequence of networks where it trains each neuron one by one with back propagation. In intelligent transport systems, RNN is crucial to process the sequential traffic data for traffic prediction, route prediction and monitoring the vehicle's speed on the road. The basic structure of RNN is shown in Figure 10.17.

10.3.4 LONG SHORT-TERM MEMORY

LSTM [2, 3] is an extension of RNNs where it expands its internal memory for the storage of past computations, and it remembers the input data over a long period. It can perform all the operations as read the information, write the information and delete the information from its internal memory. The RNN suffers from the vanishing gradient point which degrades the performance of the network. Hence, the LSTM is introduced to give a suitable solution to the vanishing gradient problem.

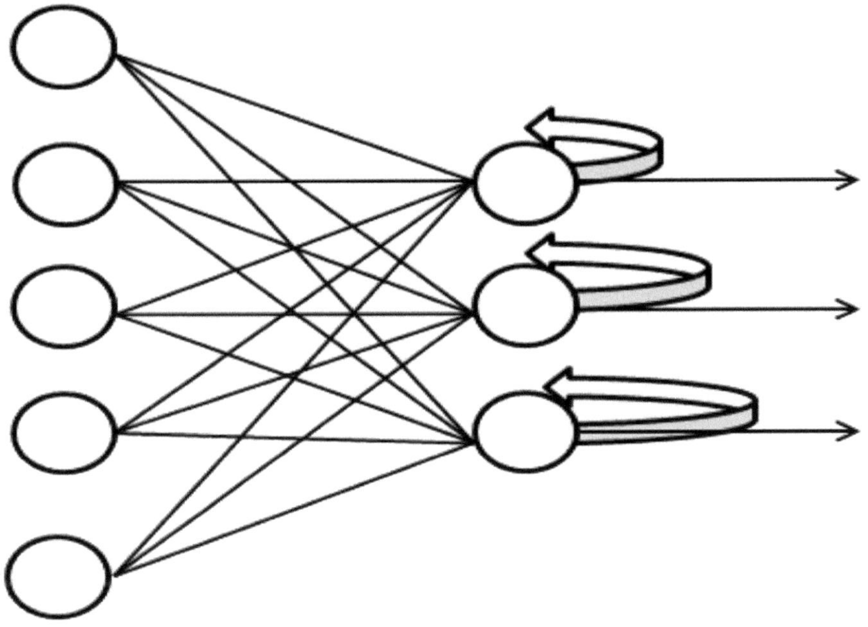

FIGURE 10.17 Structure of RNN.

In each cell of hidden layers, there are 3 gates presented in the LSTM network. These gates are used to control the information flow that is needed to find the output in the neural network. The gates are listed below:

- Input gate which determines the new input;
- Forget gate deletes the details if it is not needed;
- Output gate shows the output at the present time stamp.

In ITS, the LSTM is significant for predicting traffic flow on the road, providing traffic information service and congestion prediction.

10.3.5 BI-LSTM (BI-DIRECTIONAL LONG SHORT-TERM MEMORY)

RNN remembers the past information by incorporating hidden layers in traditional neural networks. For time series problems Bi-LSTM (bi-directional long short-term memory) is very useful. It predicts the next output based on the past inputs. It achieves it by involving gates in the network system. In Bi-LSTM, two independent RNNs are connecting together to form a network. In this way, it remembers the forward as well as backward information about the given input at each time step (Figure 10.18).

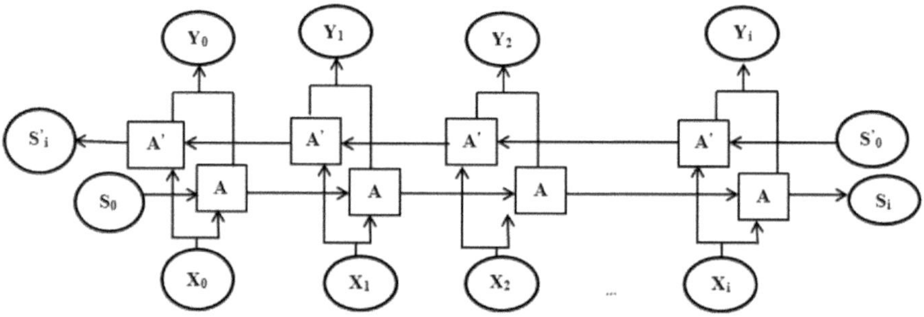

FIGURE 10.18 Bi-LSTM network.

The foundation of Bi-LSTM is long short-term memory (LSTM) which consists of 3 different gates known as input gate, output gate and forget gate. It uses not only the candidate values but activation values for processing the input values. It produces two output values such as activation value and candidate value.

A candidate value is calculated by:

$$C^{N} = \tanh(W_c\left[a^{<t-1>}, x^{<t>}\right] + b_c)$$

where; W is the weight values for the input; a is the activation; x is the input; and b bias.

The output of the Bi-LSTM network $\hat{y}$ at the given time t is obtained from both activations know as forward and backward as:

$$\hat{y}^{<t>} = g(W_y\left[\vec{a}^{<t>}, \overleftarrow{a}^{<t>}\right] + b_y)$$

In this equation:

$$\vec{a}^{<t>} \rightarrow \textit{Forward Activation}$$

$$\overleftarrow{a}^{<t>} \rightarrow \textit{Backward Activation}$$

As Bi-LSTM has two recurrent units it learns the past information as well as future information to predict the output. Bi-LSTM is one of the emerging techniques for many ML problems. This technique to be applied for various applications of STS.

10.3.6 GRU (GATED RECURRENT UNIT)

Gated recurrent unit (GRU) is similar to LSTM as both uses gates for processing the data by controlling them. The only thing missing in GRU is cell state. It uses only hidden state. This is the reason why it is easy so easy to train GRU than LSTM. The basic model of GRU is given in Figure 10.19.

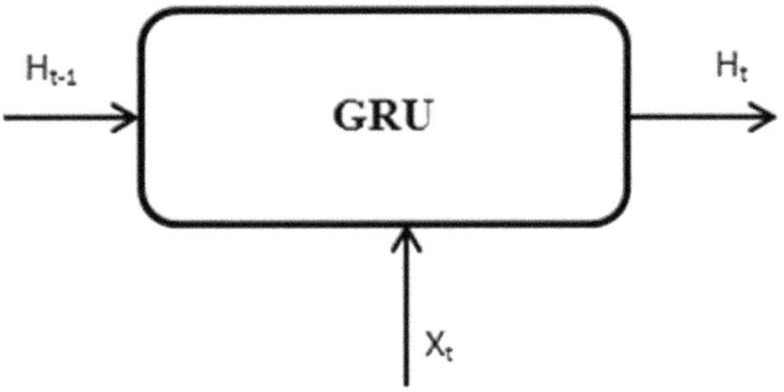

FIGURE 10.19 GRU model.

Note: X_t is the given input at timestamp t; H_{t-1} is the hidden state of timestamp t–1; and H_t is the hidden state at time stamp t. The current state $H_{t,}$ will be passed as input in the next timestamp.

GRU consists of two gates, unlike three gates in LSTM. It consists of:

1. **Reset Gate (Short Term Memory):** Reset gate acts as short-term memory for the hidden states. The equation for reset gate is given below:

$$r_t = \sigma(x_t * U_r + H_{t-1} * W_r)$$

Since the sigmoid function is used for reset gate, the value of r_t will be between 0 to 1. U_r *and* W_r are known as weight parameters.

2. **Update Gate (Long Term Memory):** Update gate is used for long term memory access in GRU. The equation is given below:

$$u_t = \sigma(x_t * U_u + H_{t-1} * W_u)$$

U_t *and* W_u are weight parameters for update gate. GRU works in two steps. The first thing is it has to find the candidate state; the second step is to find the hidden state.

The candidate state is identified by the given equation:

$$\hat{H}_t = tanh(x_t * U_g + (r_t * H_{t-1}) \circ W_g)$$

The reset gate output is multiplied by the input and hidden state of the previous time stamp and finally tanh function is applied on it. The output is hidden state of the candidate value.

10.3.6.1 HIDDEN STATE

After the candidate state the current hidden state H_t is generated. The purpose of update gate is to create a hidden state. The new information from the candidate state and historical information from the previous state are controlled by the single gate whereas in LSTM, it is done by implementing two gates. This feature makes GRU simple than any other sequence network model. The equation for calculating hidden state is given below:

$$H_t = u_t * H_{t-1} + (1 - u_t \hat{H}_t)$$

From the equation, it is known that hidden state purely depends on the past information as well as the present information. The first term in the equation points to the information from the previous state and the second term in the equation specifies the information from the current state. So, the value of u_t is near to 0, then the hidden state will purely depend on the current state and not on the historical information. In this case, the second part will be 1, hence the hidden state relies on the information from the candidate state only. It is clear that the value of u_t is so crucial in calculating the hidden state [24].

10.3.7 ATTENTION-BASED GRU

Initially attention mechanism in neural networks is introduced for encode-decoder based architecture for handling natural language processing (NLP) tasks.

Later it has been extended to other ML tasks to increase the learning capability of the system. Distance between the input sequences will not affect the dependencies while using attention mechanism. In Figure 10.20, Bi-LSTM model is used with attention mechanism. This technique can be incorporated with any other neural networks.

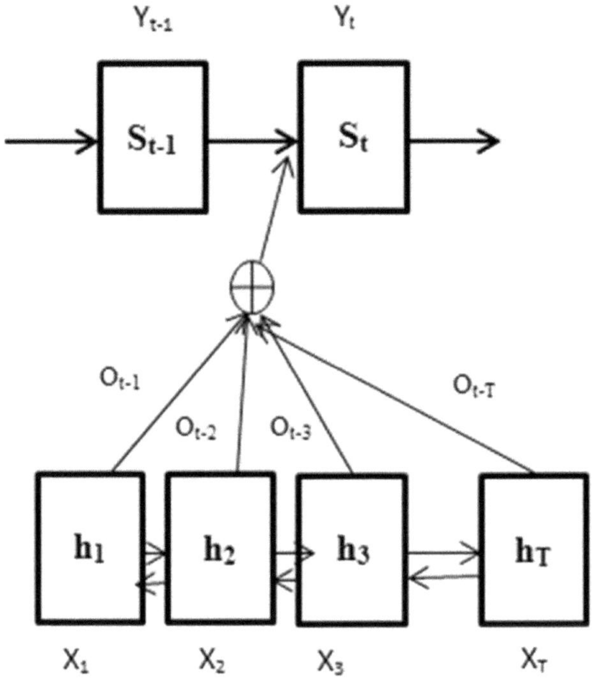

FIGURE 10.20 Attention mechanism.

Note: S_{t-1} is the previous input sequence; S_t is the current input sequence; h is the hidden state; and X is the current input.

The hidden state vectors represent the input and output sequences.

$$h_j = \left[\vec{h}_j^T ; \overleftarrow{h}_j^T \right]^T$$

The vectors h_1, h_2, …, h_T represents the number of words in the input sequence. Here the weights are calculated by feed forward networks. The context vector C_i is generated by the weights obtained from the sum of annotations.

$$C_i = \sum_{j=1}^{T_x} \alpha_{ij} h_j$$

SoftMax function is used to create weights α_{ij}. Softmax function represents a function which converts the vector of x-values into a vector of x-values whose values will be summed upto 1. The input values can be positive number, negative number or even more than the value of 1. But after the SoftMax function is applied to the values, it would be changed to the value between 0 and 1. Hence it will be in terms of probabilities.

$$\alpha_{ij} = \frac{\exp(e_{ij})}{\sum_{k=1}^{T_x} \exp(e_{ik})}$$

The above equation is used to calculate the weights of the network.

$$e_{ij} = \mathrm{a}(S_{i-1}, h_j)$$

e_{ij} represents the output score of feed forward network. Thus, attention mechanism boosts up the learning capability of the neural network.

These are the some of the emerging technologies which are applied for STS to handle all the transportation issues. The system reads the data from the various sensors using IoT technologies. After collecting the data from the sensors, various ML methods which are discussed above are applied, and the decision is made. These technologies bring the intelligent transportation system.

10.4 SUMMARY

In this chapter, we discussed various artificial techniques for giving smart solutions for the transport system. Initially we discussed about the existing transportation system and their problems. Then in the second section, we discussed the different kinds of AI technologies which deal with smart transportation. The last section discusses some of the new AI technologies which handle the future smart transportation services. Now a days, deep learning techniques gives solutions for all sort of classification problems. Hence various deep learning techniques such as RNN, BiLSTM, GRU can be used to provide solution for STS. To increase the accuracy of the system, the model's parameters can be tuned manually or automatically.

KEYWORDS

- **artificial intelligence**
- **congestion**
- **genetic algorithm**
- **intelligent transport system**
- **pollution**
- **radio frequency identification**
- **smart transportation system**
- **traffic control**

REFERENCES

1. Sadeghi-Niaraki, A., Parima, M., Maryam, S., & Soo-Mi, C., (2020). Short-term traffic flow prediction using the modified Elman recurrent neural network optimized through a genetic algorithm. *IEEE Access, 8*, 217526–217540.
2. Navin, R., Sovit, B., Hong, P. Z., Hoon, K., & Pervez, K., (2020). City-wide traffic congestion prediction based on CNN, LSTM and transpose CNN. *IEEE Access, 8*, 81606–81620.
3. Miaomiao, C., Victor, O. K. L., & Vincent, W. S. C., (2020). A CNN-LSTM model for traffic speed prediction. *IEEE 91st Vehicular Technology Conference (VTC2020-Spring)*.
4. Zhihan, L., Ranran, L., & Amit, K. S., (2020). AI empowered communication systems for intelligent transportation systems. *IEEE Transactions on Intelligent Transportation Systems*.
5. Nawaf, O. A., Ahmad, F. K., & Aws, M., (2019). Intelligent transportation and control systems using data mining and machine learning techniques: A comprehensive study. *IEEE Access, 07*, 49830–49857.
6. Ali, A. M., Kamarudin, A., Ahmed, M. M., & Deprizon, S., (2019). *Expert System in Engineering Transportation: A Review, 14*(1), 229–252.
7. Priyadarshini, P., & Sangeeta, S., (2016). A review paper on intelligence transportation system. *International Research Journal of Engineering and Technology, 3*(12), 1161–1164.
8. Li, J., Dridi, M., & El-Moudni, A., (2016). A cooperative traffic control for the vehicles in the intersection based on the genetic algorithm. In: *4th IEEE International Colloquium on Information Science and Technology (CiSt)* (pp. 627–632).
9. Wanxia, Y., Lina, L., & Weicun, Z., (2015). Traffic prediction method based on RBF neural network with improved artificial bee colony algorithm. *International Conference on Intelligent Networks and Intelligent Systems*.

10. Jiaxu, Z., Zhide, C., & Yali, Z., (2015). Intelligent traffic light model based on Grey-Markov model and improved ant colony optimization for dynamic route guidance. *IEEE 2nd International Conference on Cyber Security and Cloud Computing.*
11. Attar, A., Chaudhary, P., Dhuri, P., Vaidya, K., & Venkatesan, N., (2015). Intelligent traffic management system. *International Education and Research Journal, 1*(4), 1–3.
12. Hossain, M. S., Sinha, H., & Mustafa, R., (2015). A belief rule based expert system to control traffic signals under uncertainty. In: *1st International Conference on Computer and Information Engineering (ICCIE)* (pp. 83–86).
13. Yeh, Y. C., & Tsai, M. S., (2015). Development of a genetic algorithm based electric vehicle charging coordination on distribution networks. In: *IEEE Congress on Evolutionary Computation (CEC)* (pp. 283–290).
14. Rajabifard, A., Thompson, R. G., & Chen, Y., (2015). An intelligent disaster decision support system for increasing the sustainability of transport networks. *Natural Resources Forum, 39*(2), 83–96.
15. Zhu, J. Z., Cao, J. X., & Zhu, Y., (2014). Traffic volume forecasting based on radial basis function neural network with the consideration of traffic flows at the adjacent intersections. *Transportation Research Part C: Emerging Technologies, 47*, 139–154.
16. Falamarzi, A., Borhan, M. N., & Rahmat, R. A. O., (2014). Developing a web-based advisory expert system for implementing traffic calming strategies. *The Scientific World Journal, 2014*, 1–16.
17. Fogue, M., Garrido, P., Martinez, F. J., Cano, J. C., Calafate, C. T., & Manzoni, P., (2013). A novel approach for traffic accidents sanitary resource allocation based on multi-objective genetic algorithms. *Expert Systems with Applications, 40*(1), 323–336.
18. Binjammaz, T., Al-Bayatti, A., & Al-Hargan, A., (2013). GPS integrity monitoring for an intelligent transport system. In: *10th IEEE Workshop on Positioning Navigation and Communication (WPNC)* (pp. 1–6).
19. Amrita, S., Sahoo, G., & Sahoo, U. C., (2012). Application of fuzzy logic in transport planning. *International Journal on Soft Computing, 03*(2), 1–21.
20. Shandiz, H. T., Khosravi, M., & Doaee, M., (2009). Intelligent transport system based on genetic algorithm. *World Applied Sciences Journal, 6*(7), 908–913.
21. Chu-Hsing, L., Jui-Ling, Y., Jung-Chun, L., & Chia-Jen, L., (2008). Genetic algorithm for shortest driving time in intelligent transportation systems. *International Conference on Multimedia and Ubiquitous Engineering.*
22. Naranjo, J. E., Sotelo, M. A., Gonzalez, C., Garcia, R., & De Pedro, T., (2007). Using fuzzy logic in automated vehicle control. *IEEE Intelligent Systems, 22*(1), 36–45.
23. Yang, Y., Hu, J., & Chen, D., (2007). Research on driving knowledge expert system of distributed vehicle driving simulator. In: *11th International Conference on Computer Supported Cooperative Work in Design (CSCWD)* (pp. 693–697).
24. Delen, D., Sharda, R., & Bessonov, M., (2006). Identifying significant predictors of injury severity in traffic accidents using a series of artificial neural networks. *Accident Analysis & Prevention, 38*(3), 434–444.

CHAPTER 11

AI-Based Green Transportation: A Sustainable Approach

KABEER TIJJANI SALEH[1], AUWAL ALHASSAN MUSA[2,4], SALIM IDRIS MALAMI[3,4], YASEMIN SARIKAYA LEVENT[1,5], and SHASHIVENDRA DULAWAT[2]

[1]*Akdeniz University, Antalya, Turkey*

[2]*Civil Engineering Department, Mewar University, Chittorgarh, Rajasthan, India*

[3]*School of Energy, Geo-Science, Infrastructure and Society, Institute for Sustainable Built Environment, Heriot-Watt University, Edinburgh, United Kingdom*

[4]*Civil Engineering Department, Kano University of Science and Technology, Wudil, Kano, Nigeria*

[5]*City and Regional Planning Department, Mersin University, Turkey*

ABSTRACT

Current urban societies are characterized by having been polluted on a daily basis due to increase in urbanization activities, industrialization, and transportation. In an attempt of having clean, healthy, sustainable, and pollution free urban environment, the concept of green buildings, green transportation or rather sustainable transportation is discussed in this chapter. Green transportation unveils the ability of the available transportation modes in our environment to produce a renewable source of energy that is

Artificial Intelligence for Future Intelligent Transportation: Smarter and Greener Infrastructure Design. Rajesh Kumar Dhanaraj, Nilayam Kumar Kamila, Subhendu Kumar Pani, Balamurugan Balusamy, & Vani Rajasekar (Eds.)

clean, defendant, and has no harmful effect to the inhabitants. This green transportation aimed in achieving sustainability in urban transportation, urban climate, social lifestyle and environmental management in general through artificial intelligence (AI) sustenance. AI is one of the components of green transportation as it refers to the recreation and application of human intelligence in machines that will be programmed and have the ability to execute task in simpler, quicker, sustainable with little or no detrimental effect to the urban environment. In this instance, the application of such intelligence in green transportation is due considered in this chapter, the role of AI in green transportation as well as challenges associated with it. While a case study of Istanbul Metropolis of Turkey was also featured.

11.1 INTRODUCTION

Long ago it's through the utilization of transportation system that societies use to share their ideas, culture, and commercializing their products. Therefore, its vividly proven that, transportation is tied to our daily basic activities. Perhaps, it's part of our culture anciently. Most of the current world developed cities rely on integrated transportation systems. While developing cities are fast growing and they cannot be free from diverse associated challenges hindering their sustainability and causing hardship to their dwellers. These challenges include air and water pollutions, soil degradation and their health and financial effects which were known to be originated form the excessive usage of fossil energy available which most conventional means of transportation is utilizing [1].

Urban environments are facing persistent problems related to pollution that expose it to different challenges. A recent study done by [2] proposes that, globally, 25% of urban ambient air pollution from Fine particulate matter (PM2.5) is being contributed to the urban environment from combustion and burning of fossil fuel by traffic and transportation activities. In this case, it should be clearly understood that road transportation contributes a significant percentage of urban environmental pollution in our cities and towns. While in most cases, air pollution causes serious problems like excessive increase of carbon monoxide (CO), global warming, ozone depletion, and likes.

As a result of the current high rise in urbanization activities, in a report by World Urbanization Prospect [3], stated that 54% of the world's population living in an urban area and this figure is expected to increase to 66% by 2050. The universal increase in the urbanization track is predictable that the global population will reach up to 8 billion by the year 2025 and 9 billion by 2040. Hence, this increase in universal growth is clearly considered to be occurring in urban areas whereas the total rural population is projected to remain stagnant at about 3.3 billion by 2035 and then drop slightly to 3.2 billion by 2050. All future population growth is thus urban population growth [4]. These developments are associated with the increase of vehicles ownership. Currently, world houses about 1.2 billion working vehicles (light weight vehicles, motorcycles, freight vehicles and public transport vehicles). These estimates were projected to reach 3.1 billion by the year 2050 as [5] report indicates. Clear indication has shown that a larger percentage of the above projections will be recorded from undeveloped and developing countries, because developed cities are making attempts in diversifying and encouraging green transportation missions and controlling their population.

According to World Energy Scenarios [6], the contribution that transportation made through the emission of carbon dioxide gas is up to 30% of the world total. Also, it was claimed by [7] that, transport activities are the second largest apart from energy sector in greenhouse gas (GHG) emission which contributes about 25% of the total GHG emission in the entire Europe. This amount is increasing at alarming rate which if not controlled it might cause destructive damage to the remaining ecosystem of the continent. Transportation sector is becoming larger and faster in energy consuming, despite these proven statistical data on how the transportation activities pollutes the environment through utilization of recognizable size of world's energy in form of fossil fuel. Unfortunately, it attracts less attention of policy makers on technical means of mitigating this energy consumption. This was due to non-availability of reliable and alternative sources of non-fossil energy sources (other sources than coal, hydro, natural gas, nuclear, solar, and wind energy sources), this has become a problem as a result of over 90% of transport activities depend on fossil energy source [8].

The ideas of green transportation are encouraging to replace the current transportation vehicles on which most of them rely solely on fossil fuels as their source of energy. These fossil fuels emit considerable amount of toxic gases (carbon dioxide, CO, nitrogen oxide and other partially burnt heavy hydrocarbons (HCs)), where 23% of these toxic gases are contributed by transport sector [9]. It was discovered by Roberts & Patterson [10] that biofuel is a relatively good replacement of conventional fuels used for transport vehicles, but it has side consequences upon adoption as its major threat to food security. Therefore, these biofuels might not serve well in the nearest future.

Despite increasing proposed works on green transportation from the last two decades still no agreeable sustainable transportation systems exist between all policies makers among countries. This was resulted from the different cultural setup, environmental laws and unique nature of living standards of the societies. These hurdles have not distracted the intention of researchers aiming at developing and implementing number of marvelous green transportation policies, of course an appreciable progress was recorded. Hence, a study on green transportation, sustainable transporting or ecofriendly transportation will be just a place to start.

11.2 CONCEPTUALIZING SUSTAINABLE TRANSPORTATION SYSTEMS

It's a common goal for every region of the world to provide moderate energy consuming, efficient, environmental-friendly, and reliable transportation systems [11]. Developing the idea of integrating the Artificial Intelligence (AI) with generalized transportation activities being it design, analysis, construction, operation or maintenance is perhaps a key point to be needed in the current fast riding digital world. Therefore, to have consistent and continuous economic, social, and secured pollution-free environment, a deeper comprehension on how the AI could be implementable to impact into the greener transportation is prerequisite in this century. However, to have a broader understanding about the concept, there is a need to clearly have an idea about the concept of green transportation and its relationship with the sustainability and environment in general.

Sustainability is the process of developing systems that is capable of meeting the current societal demand without hindering the future basic need that will be expected from the outcome delivered by those systems [12]. The above definition is likely the most reliable and accepted definition of sustainability as it covers up visible and virtual differences such as social, economic, environmental, cultural set up, beliefs, and ethics of the societies [13]. Although this was argued that it has cover extensive broad of living which is merely complex to be adhered with. Nonetheless, at least it has captured the three triple bottom line elements of sustainable development [14]. However, considering social, economic, and environmental impact of the sustainability to the society, with this definition it can be precisely shown that sustainability cannot be complete without proper and careful consideration of transportation activities.

While Green environment as it was defined by [15] is the combination of natural and partially manmade or modified natural environment designed, control, and managed to offer uninterrupted similar services offered my naturally made ecosystem such disaster risk management, water, and air pollution mitigation, control of climate change. In addition, it includes other unique features such as terrestrial, marine areas, etc. The main aim of this combination is to provide and facilitate an enhanced quality environment to the inhabitants, serves as a barrier to biodiversity degradation and helps the ecosystem in providing social, economic, and environmental benefits. The green city or sustainable city is designed to have standard, safer, secured, and non-polluted surrounding, with additional advantage of minimal accident occurrence. Its unarguable that sustainable transport is prerequisite to a green city development as it plays a vital role to any developed city in the world. Therefore, proper integration of sustainable transport to other sector forming the green city is necessary. This synergetic approach is to be understood holistically through public-private partnership.

Additionally, green transportation may have a clear and better understanding as another term used interchangeably with sustainable transportation which refers to all transportation systems that has low negative impact of pollution emission to the environment, maintaining optimum fuel consumption, with proper considerable balance between inhabitants and environments. It also, includes a well-planned and constrained-free forecasted future of the cities based on their master plan which encompasses practices and policies, in term of economic viability, social equitability, and benefiting the environmental consistently with long term future plan without leaving any transportation gap [16]. Green/sustainable transportation system generally shows emphases on societal planning, modern technology and politics. And its main objective is to ensure an effective distribution and transit of goods, people, and the implementation of high-quality transportation services in urban environments [17].

Green transportation can be viewed in two broadly perspectives either as a means of transportation, which includes walking, bicycle, regular public transport and rail transport, or from the second perspective which consider vehicles of transportation, this includes hybrid-energy vehicle, natural gas vehicle, electric vehicle, hydrogen power and solar powered vehicles [18]. The hierarchy of green transportation considering the first category was initially proposed by [19] in which they categorizes it by walking as the most effective means of attaining greener transportation, followed by bicycle, public mass transit, shared vehicles and last of this order single driving vehicles. Well integrated transportation systems tend to contribute immensely to the health, education, employments, and economic developments but it also contributes toward rapid increase of greenhouse effect due to these large fuel consumptions [20]. Its worthy to note that in this work, the sustainable transportation or green transportation as two terms were used interchangeably to relate it with sustainability (Figure 11.1).

FIGURE 11.1 A typical prototype of green environment.

Source: Adopted from [21].

As most of the technical systems are having their independent features and rules likewise, sustainable transportation is also related inclusively. These rules were established by [22] and they are as follows:

1. **Reduced Travel Needs:** This has to do with establishing a relationship between information communication technology and transportation such as supporting the idea of online bus booking, online shopping, etc.
2. **Transport Policy Shift:** This has to do with Paradigm shifts to new strategies by replacing the excessive use of cars with walking

and cyclic means of movements. In this regard walking and cyclic means of transportation should be given priority especially in urban areas followed by public transport and least with conventional transport system which car usage is dominant.

3. **Distance Reduction:** This has to do with planning new cities roadways where an accessibility and mobility will be provided to ease the difficulties for the road users in accessing the cities services (housing, commerce, recreation, and education facilities). This rule encourages the establishment of green urban island where it's declared as no car free zone.
4. **Integration of Artificial Intelligence (AI) to the System:** This has to do with the adopting of new technologies in vehicles manufacturing, alternatives fuels, increase in renewable energy sources, timely accessibility to major cities centers, etc. This rules also includes action toward rational drivers' behaviors, strict traffic rules and eco-friendly vehicles provision.

11.3 WHY GREEN TRANSPORTATION?

Apart from the contribution offered by transportation activities for the rapid urban development recorded by most of the developed nations, these activities also contributed immensely toward other challenges associated with the urban development. These challenges include relatively high traffic growth, increased nuisance of traffic in high populated area, excessive release of converted fossil energy into poisonous gases to the environments, which in turn causes air pollution, increasing travel time of the commuters while at the same time increasing the energy usage and persistent environmental degradation. Among, these are the few challenges that sustainable transportation are targeted to solve. The solution to above-mentioned challenges can be put into practice through the adaptation of proper and coherent environmental assessment, optimistic, and resilient infrastructural financing and minimizing the amount of CO_2, through encouraging the usage of other alternative means of transportation [23].

Unarguably, transportation systems are socio-technical in nature, thus dealing with such non-autonomous, complex, and diverse system is merely difficult considering the number of factors having featured variabilities it houses [24]. For this reason, some regions have already embarked on proper sensitizing local communities on the consequences of misusing these systems in accordance with their initial purposes [25]. Therefore, proper implementation of green transportation contributes not only to lower the carbon emission and provide environmental-friendly means of transportation, but also aids in promoting health through increasing the level of air quality and decreasing the amount of energy consumption.

Its worthy to note that millions of lives were claimed by road accidents annually across the world, most of were caused by cars. With this figure in 2007, China has established and implemented prompt decision by providing transit-oriented development (TOD), a scheme which discourage the persistent use of cars as the most commonly used means of mobility in urban areas by attracting communities toward a mass transit aiming at achieving high quality life [26]. The scheme has impacted immensely toward achieving the intended goal of green transportation development. Therefore, encouraging the adaptation of sustainable transportation will fragmentally reduce the occurrence of such menace to a minimal level.

As the second largest contributor of the green gas emission, transportation contributes negatively to the environment [27]. Decoupling transportation and CO_2 is necessary through possible mitigation measures such as employing consistent sustainable measures [28]. Of course, considering other parameters of sustainable development, the rate of implementing green transportation is slower with respect to reducing the level of toxic gases emissions [29]. Regardless of all the efforts made in attaining most the sustainable goals of quality of life, basic freedom to all, accessible education by all, food security, good health care, other social capabilities, but sustainable transportation has less attention from the policies maker's despite being one of the key drivers to the nation's development. Therefore, special and independent care needs to be imposed on sustainable transportation. This is because human mobility and development are two non-separable phenomena, and at the same if not controlled to a certain level, over mobility will serve as destruction to the development [30].

The critical argument about the components of sustainable development by the participants of [31], it was concluded that a special attention should be drawn toward transportation and mobility as they gear up and serve as the central system of sustainable development. They also provide linkage of urban-rural areas, enhancing the level of productivity and provide economic and social improvement. Therefore, during that historic conference, it was agreed a new policymaking system should be adopted at every level of governance to adhere with the conference recommendations on sustainable transportation [32].

11.4 SETBACKS OF CONVENTIONAL TRANSPORT SYSTEMS

All the tried attempts were meant to protect and improve the environment from the negative impacts of transportation. These impacts have been categorized into three by [33]:

1. **Direct Impacts:** These are the main effects that could easily be seen and noticed, and understood such as noise sound and CO emission.
2. **Indirect Impacts:** The consequences of these are more likely not to be understood easily, they may have direct link to respiratory problem and other cardiovascular diseases such effects are mostly caused due to incomplete gaseous combustion of internal combustion engine.
3. **Cumulative or Tertiary Impacts:** These are having direct effects to the climate change, global warming and other natural consequences such as carbon dioxide emissions.

On boarder term, below are some of the challenges from conventional transportation systems:

1. **Reducing the Environmental Air Quality:** Poisonous gases released by the vehicles affect the human health and natural build environment, more than a century since the inception of formal transportation systems into existence. Respiratory related diseases are escalating such as asthma, bronchitis, heart attacks, etc., were claimed to be associated with transport emitted gas which was epidemiologically reported by [34], these respiratory related diseases lead to the death of more than half of million people annually. Also report by [35] claimed that an approximate death of about 2 million is recorded across the world due to the air pollution diseases. Apart from the aforementioned effects, the excessive usage of fossil fuels as a source of energy causes depletion of natural resources, global warming, and environmental degradation.
2. **Noise and Vibration:** Some part of the human nervous system is affected densely by excessive noise. Among problems caused

by the noise and vibration from vehicles operating sound, horns, and other sounds associated with vehicles are pain and hearing fatigue, sleep disturbance, cardiovascular effects, immune system impairment, etc. It also affects underage speedy and steady leaning assimilation [36].

3. **Accident Occurrence:** It was estimated that over 1.2 million lives are usually accumulated by accident annually across the world, with more than 50 million injuries. These estimates are mostly accounted for in undeveloped and developing countries [37]. It was projected the road accident will be the fifth leading cause of death by the year 2030 [38].
4. **Global Climate Change:** As it was mentioned before, more than a quarter of the fossil-based carbon dioxide released to the atmosphere was sourced from vehicles. In 2009, transport was responsible for 23% of global carbon dioxide emissions from fossil-fuel-based sources, while the energy sector accounted for 41% of these emissions [39]. However, by 2035, transport is expected to become the single largest GHG emitter, accounting for 46% of global GHG emissions [40].
5. **Disruption of Naturally Build Environment:** Road and other related transport systems construction is associated with the depletion naturally build environment.
6. **Traffic Congestion:** It is inevitable to the delay of the general activities of the road users. It was estimated by [41] that as for 2010, United Stated of America losses about $101 billion as result of traffic congestion majorly Resulted from the fuel wasted during the delay and time spend by passengers. Also, Thailand's GDP was shrinkage by about 6% as a result of congestion experienced on the Bangkok city's major roadways [42]. According to the research conducted by [43], the lost cause by congestion in China cities constitute about 2% of its GDP.
7. **Fossil Energy Security:** Continuous usage of petrol-based fuels by a particular country may lead to the perishing of its natural resource which in the long run might expose the nation to the energy-insecurity stage. While to the countries relying on importing fuel related product may be threatened with price hike and invariably impacts and experienced shortfalls on their national treasury.

11.5 POSSIBLE SOLUTIONS

To curb these aforementioned challenges, many developed cities have introduced technological leapfrogging, i.e., the idea of sustainable transportation through series programs encouraging the usage of alternative means of transportation which include electric cars, improving the usage of commercial buses as replacement of commonly private cars which intended to have high fuel consumption cost maintenance cost as well as contributes in polluting the environment. Also [44] claims that for a city to implement the idea of green transportation it has to follow the series of activities without skip these are: basic access stage – (provision of local/access road) paratransit dominant stage (availability of public transportation), early motorization stage (provision of means of transportation means), full motorization stage (availability of public and private transportation means), quality of life stage (integrated and effective public transport systems). However, [45] stated that since the 1992 Earth Summit in Rio, when Agenda 21 which sustainable transportation is inclusive and many cities such as Amsterdam (Netherlands), Bogota (Colombia), Brisbane (Australia), Copenhagen (Denmark), Curitiba (Brazil), Guangzhou (People's Republic of China (PRC), Portland (United States), and Seoul (Republic of Korea), have embraced implementing the idea.

11.5.1 ELECTRIC VEHICLES (EVS)

The most challenging and accepted fact in the nearest future is that the world oil reservation will keep declining, while the rate of its consumption will comparatively rise as a result of factors earlier mentioned. Therefore, to avoid necessary disruption of transportation sector which heavily rely on the oil production, its requisite to implement other innovative technological advancements to replace the over dependence on oil as the source of its energy. This can be achieved through the introduction and adaptation of proposed green transportation schemes such as electric and hybrid vehicles as it reported by [46]. These vehicles (electric vehicles, hybrid electric vehicles, plug-in hybrid electric vehicles) are capable of minimizing the negative effects caused by the conventional vehicles [21]. Although, the idea of EVs has faced many challenges from its inception to date as many developed and developing nations in collaboration with well-developed manufacturing companies such as spaceX (TESLA) in USA, Nissan (LEAF) in Hong Kong Tata in India, etc., are still struggling to be accepted by the societies.

In order to minimize these challenges faced by EVs, the Hong Kong government has installed about 300 charging points across 18 districts. While China along as the leader of the electric vehicle production in the world are having more than half of million EVs across the country. The Nissan leaf product as the first type of electric vehicle can travel at upper range speed of 145 kM/hr. and the charging can last for 160 KM. It is not astonishing that the production of EVs was dated back to 1838 before the inception of combustion engine [47]. The comfortability and massive commercialization of these combustion engine in later years brought about the drop of the EVs. Afterward, several attempts have been made to reintroduce the EVs which most of these attempts failed due to many factors. But based on the recent progress and support could made more than 50% of the total sales by the year 2050. Making up to 25% of the world vehicles sales (Figure 11.2).

FIGURE 11.2 Typical charging process of electric car.
Source: Adopted from [48].

Another contribution made by The International Renewable Energy Agency (IRENA) [49] as an advancement to the green transportation are the recommendations of implementing EVs and adoption of biofuels. This campaigned recommendation has encouraged the usage of EVs at the end of 2016 to more than 2 million across the world. Again, much is also needed to speed up the process. With regard to biofuels, several research have shown that they will serve the same purpose with the conventional energy sources (gasoline, diesel, and kerosene) available to internal combustion engines. The International Renewable Energy Agency [49] suggested that upon adopted at least 10% of the energy exhausted by transportation sector will be saved by the year 2030. However, it has negative consequences as it was mentioned in the beginning of this work.

11.5.2 CAR FREE ZONE

The idea of care free-zone has been adopted and been in practice in some developing cities such as Buenos Aires (Argentina), Cartagena (Colombia), Curitiba (Brazil), Old Delhi (India), Quito (Ecuador), Sao Paulo (Brazil), and Shanghai (PRC) have permanent car-free areas, as another stage to improve and support the idea of green transportation.

11.5.3 GREENWAYS

A recent development aiming at replacing the conventional, dominated, and motorized landscape with green plantation in most pedestrianized runways was greenways, apart from its initial purpose of improving the green transportation system, they also serve as recreational zones in cities.

11.5.4 IMPROVING PUBLIC TRANSPORT

It's difficult to reduce the usage level of conventional means of transportation through alternative means such as massive deployment of public transportation, bicycling, walking. Despite all the challenges associated with these means it impacts less on the environment based on pollution reduction, fuel costs, economic growth and building strong social networking. By the way considering the active implementation of overall green transport integration in different forms to transport services, it will improve the attractiveness and convenience of public transport. Having implemented the idea of green transportation, it is also necessary to consider remodeling the current system of public transport in order to attract public attention to it and bridging the competition of the private car owner ship among the commuters.

Urban road-based strategies can be implemented as proposed strategies: they include bus rapid transit (BRT) paratransit upgrades, urban rail-based, these includes streets trams, at grade light rail transit (LRT) elevated light rail transit (ELRT), metro, proper consideration of fare system which can be done by online booking, as well as keening to promote the use of smart ticketing which allows passengers to move seamlessly between different modes [50]. Another means of attracting public attention to the less energy-consumption means of transportation is that governments must be ready to modify public transport systems by increasing the level of service, proper parking facilities provision, an affordable fare to the commuters. Provision of integrated routes network purposely for public transportation, restructuring the stations in term of comfortability, security, conveniences, optimizing the taxation system, and provision of terminals along major crowded places such as market for simple loading and unloading goods and services, as well as integration of these routes and terminals with freight vehicles routes, feeder service and other related mass transit systems such linkage to airports, seaports, and rail terminals needed for urban development [51].

11.5.5 SMART CITIES ESTABLISHMENTS

Although the entire sustainable transportation system is part of the smart cities agenda. But nonetheless rapid evolvement of the smart cities will jointly provide assured means of proper green transport systems.

11.5.6 ECO-FRIENDLY URBAN SCHEMES

Most of the developed cities in the world are threatened with the large volume of toxic gaseous emission from both mobile and static sources. Therefore, an advanced means of reducing such menace is by encouraging eco-friendly innovation. These includes provision of cyclic tracks, BRT system (BRT) intelligent transport system, urban freight management, hybridization of fossil fuel and non-fossils fuel vehicles.

11.5.7 GREEN TRAVELING AWARENESS AND ADVOCATES

The already implemented active green transportation in use by most of the cities in the world lacks public participation despite the support offered by the government. Therefore, active promotion of green traveling through adopting functional sensitization programs which can be implemented by the support of schools, local communities, advocates groups, environmental protection organizations, transportation actors, and voluntary organizations solely for green transportation. These programs should obtain full support from the government through adopting flexible administrative, economic, and technological approaches and further integration of the system with guided and updated scientific research. This advocate can be embedded with prompt initiatives from NGOs and collaboration with all the relevant bodies [52].

11.5.8 PROVISION OF LONG-TERM RESILIENCE SYSTEM

The problems associated with the most models developed is non-resiliencies after a short period of initial implementation. Proper design of long-term sustained mechanisms will help green transportation to be part of the public culture. This can be achieved by adopting intelligent transport system, integrated road networks, accessible mobility, policies on energy conservations and environmental protection [18]. In addition, a well-planned and oriented-transit and zoning, developing reliable policies that favor the land use, development of land belt and urban boundaries will assist in longing the scheme.

11.5.9 TELE-MOBILITY

Avoidance or reducing of unnecessary trips using new paradigms of implementing green transportation is another way in which the use of information technology as a replacement of some manageable activities through the adoption of TOD. This focused on adopting of these sophisticated components for tele-mobility which include telecommuting, telework, teleconferencing, teleshopping, and telemedicine. In broader term the tele-mobility is component of AI which discourage excessive travel demand [53].

11.5.10 RESTORATION OF CLIMATE

The sustainable transportation system contributes immensely toward improving the cleaner environment. Much attention has been given to this sector recently by most of the countries through public-private partnership, which has now become inevitable to the economic and social development. Transportation sector can be supported by the implementation of policies, appraisal, and managerial commitment by the parties responsible.

The global warming is far beyond the individual or community problem but rather world ones. Although it can be sub divided into different levels depending on the laws and regional cultural setup. Among these initiatives were global fuel economy initiatives (GFEI) [54]. This was initiated to control and improve the fuel consumption management, one of their recent reports claims that carbon dioxide will drastically reduce to half in the next three decades, while more than 5 million barrels of oil could be saved annually. The (GFEI) was formed by collaboration between united nation environment program [55] and international transport forum and federation international automobile foundation.

1. **The Mobility Model Partnership:** Popularly known as MoMo, this model was developed to have a regional-based database of world transport, the model contains mode of transportation, fuel consumption, regions, previous record of transportation of the region as input variables and provides the next decade's prediction of energy consumed by transportation activities based on quantitative analysis using these parameters. The successful implementation of this model has brought about future forecast of energy use, vehicle, and fuel cost, as well as GHG emission.
2. **The Electric Vehicles (EVs) Initiative:** Although it was initiated long before by different countries through the public-private partnership, it has experienced set back from some countries due to the lack of enabling environment.
3. **Sustainable and Low Carbon Transport Partnership:** The drawback of implementing the most sustainable transportation policies is recorded and experienced by developing countries. Due to this problem, this partnership was established to serve as sensitization and knowledge sharing medium to these countries affected. This partnership includes more than 50 organizations amongst are

United nation, NGOs, research institutions, development banks and private agencies.

4. **Urban Sustainability Framework:** This was initially developed by [56] as the strategized transportation planning measures focused on evaluating the green transport performance based on the triple bottom line of sustainability (economy viability, social equitability and environmental maintainability) to monitor the green transportation track using household, market access point, land use and local road are parametric input measures.

11.6 THE IMPACT OF IMPLEMENTING GREEN TRANSPORTATION TO SOCIO-ECONOMIC DEVELOPMENT

In consideration of economic benefits, green transportation hugely permits savings and reduce the construction and maintenance cost due to its borrowed natural nature, for instance, storm run-off that vacates from road surface usually pollutes water bodies, causing health and financial distress to inhabitants. Direct replacement of these conventional roadways and another structural component of highways with green transportation systems assists in significant reduction of this cost. In addition, porous pavement was found to reduce almost 25% of the cost, if it is replaced with the conventional pavement as [57] claims. Although the live span of standard pervious pavement might last up to 25 working years if proper maintenance was adopted.

Another important point which needs to be considered offered by green transportation is the reviving socio-economic sector through creation of tourism attraction sites. Relevant design and construction methods such as frequent usage of permeable pavements which in turns reduces management cost and environmental pollution and maintaining the ecological buildup within the landscape and massively reduces the traffic congestion. Again, the entire transport systems and other infrastructures are prone to destruction by extreme weather conditions as well as unexpected natural disasters such as floods, landslides which normally cost lives and millions of dollars annually. One of the proposed solutions to reduce such destructive phenomenon was the provision of green bridges and eco tunnels [58]. Green embankments integrated along roads also serves as efficient noise barrier, help in pollution reduction through filtering the storm-water flows.

Furthermore, green construction is adopted with the integration of vegetated drainage, porous materials, permeable pavers, into streets or parking lots [59]. Therefore, integrating green transportation into sustainable environment will help in reducing carbon footprint, neutralize the effects caused by it on land, water, and other features within the environment. The provision of green transportation will also help in increasing its residences to the disaster (Figures 11.3 and 11.4).

FIGURE 11.3 Permeable pavement.
Source: Adopted from [60].

FIGURE 11.4 Green bridges and eco-tunnels.
Source: Adopted from [60].

11.7 INTRODUCING AND IMPLEMENTING THE CONCEPT OF GREENER LOGISTICS

Ahead of summing up the solution to the challenges of conventional transportation, it's vital to include the concept of green logistics, due to its functional impact to the economy and aiding in friendly environment development. Freight transportation which most of the logistics activities rely upon, contribute immensely to polluting the environment, causing harm to the reserved natural resources through depleting the fossil fuels sources [61]. Green logistics encompasses the distribution, energy consumption, manufacturing, production, and raw material mining [62]. Therefore, overlapping the Intelligent transport system to it will improve the entire activities of the logistics processes through provision of safe, less-energy consumption, and affordable systems.

11.8 THE ROLE OF ARTIFICIAL INTELLIGENCE (AI) IN GREEN/ SUSTAINABLE TRANSPORTATION

Inevitably, there is need of overlapping green transportation with AI, because the rapid growth recoded by the AI in many fields within the first two decades of 21st century reaffirms it. The sole aim of AI in sustainable transportation is to avoid the manual or human maintenances to transportation demand while at the same time improve the energy-usage efficiency through the introduction of new technology. Therefore, deployment of AI in green transportation will lead to have another aspect which termed as smart greener transportation. Many countries have embarked on developing smart cities and smart mobility, some of these technologies includes:

1. **Geographical Positioning System (GPS) Application:** Inclusion of GPS to monitor the activities of transportation within the city via the usage of sophisticated and smart devices, controlling, and management of the drivers, passengers, other road user behavior's will have enhanced the security and reduce the traffic and risk of accident on the road. While, vehicles routes, fare collection, smart taxing, can also be attained with little or no hindrance if the system is properly implemented and managed.

2. **Smart Traffic System:** The application of smart traffic system can be used to link vehicles within specific designed area for detecting and mitigating the sensed traffic congestion and averted it through alternative routes. This can be used as they can automatically present the available routes for the drivers. The idea will help in saving energy, time, and reduce the travel time, while adding efficiency to the systems by reducing the toxic gaseous emission to the environment. They will also alert drivers on the potentialities of exceeding speed limits within urban areas [63].
3. **Smart Streetlight:** These are sophisticated forms of streetlights that have a ranged regulatory that normally detect the presence of vehicles or other road users using the mounted sensors. They are usually adopted in local streets with less congestion to save energy [64].
4. **Operational Management Systems:** Another paradigm for implementation of the AI in the sustainable transportation is by adopting operational management systems to support the decision support system of transportation, these support systems can be used to smoothly analyze and provide relevant information of the systems needed such information includes traffic flow data, road accidents data, etc. Based on the data collected a proper planning can be laid. For these operational management different AI approaches such as data mining, genetic algorithm (GA), artificial neural network, evolutionary computation, situational analysis, simulation, and advanced modeling, reasoning based on precedents, retrieval knowledge from the database were employed as [65] explicitly explained. Although these operational management activities need very experienced personnel for the collection, analysis, presentation, and retrieval of data. While the adequacy and quality of the data can help immensely toward the simulation process.

11.9 CHALLENGES ASSOCIATED WITH GREEN TRANSPORTATION

For a reliable and working systems to exist it must be associated with different challenge, likewise this idea of green transportation is of no exception in this regard. Below are some challenges of sustainable transportation:

1. **Public Sensitization, Awareness, and Participation:** Hindrances and setback factors to most government policies are lacking awareness from the receiving end, therefore different awareness program should be set by government at different levels to encourage and promote public acceptance and participation in green transportation adaptations. Although, it may be difficult to some region due their lower literacy level. The level of awareness lacking between the stakeholders is another factor hindering the implementing the core aim of green transport. Therefore, knowledge sharing, and idea transmission will help in bridging the gap.
2. **Non-Availability of Smart Changing Stations:** Green vehicles and charging stations are inseparable, the abundance of charging stations and rate of charging should be considered while designing the system. In most cases the electric and hybrid vehicles will be affected much by this. Installations of such charging stations should be based on user preference for easy attraction of public into the system.
3. **Power Storage:** The power source of both electric and hybrid vehicles is battery, and these batteries are associated with some setback attributes. These are heavy and expensive power storage, longer charging period, toxicity in nature as it contains heavy element which will pollute environment if not properly disposed. Lack of detective control system for overcharging and under-charging the battery is also part of the challenges.
4. **Non-Secured Walkways Facilities:** Despite its advantage to low income earners, non-motorized transport (NMT) (walking and bicycling) transport is less safe and conveniently non-attractive. Because of less barrier to extreme high levels of noise and air pollution from the nearby motorized roads, exposed to hazardous level of high-speed traffic, lack of properly and well-designed crossing facilities, insecurity at some certain sections which leads to criminal acts such as robbery.
5. **Funding and Other Political Supports:** Lack of funding is another challenge faced by the processes of green transportation. Therefore, proper provision of necessarily needed funding which was proven to be less when compared to the conventional methods employed, and practical implementation will be among the priority concerns to all decision makers responsible for green transportation adaption.

11.10 AI-BASED SUSTAINABLE TRANSPORTATION (CASE STUDY OF ISTANBUL METROPOLIS TURKEY)

11.10.1 THE FRAMEWORK OF THE CASE STUDY

In trying to achieve the goal of this research, a questionnaire samples were established with 307 responses which were used to evaluate and measure citizen's participation in using green transportation systems with the values they seem to attach in making the environment a safer one. This study chose the city of Istanbul, Turkey as the major case study to obtain the related answers and information needed for the research. The response and results were analyzed and presented in the form of figures representing the respondent's opinions in different sections.

The questionnaire contains two major sections – in the first section, the demographic, social, and economic characteristics of the respondents were surveyed. This was done to assess the level of the social and demographic values of the respondents as a basic indicator to evaluate their need for a sustainable transportation system. While the second section, the respondents' understanding of the term green transportation, respondent quality of life, the exposure of green transportation facilities to the respondents were asked and reviewed as the indicators that can give the information of the significance of using the green form of the transportation system.

11.10.2 SECTION ONE: SOCIO-ECONOMIC BACKGROUND (TRIPLE BOTTOM LINE OF SUSTAINABLE DEVELOPMENT) (Figures 11.5–11.10)

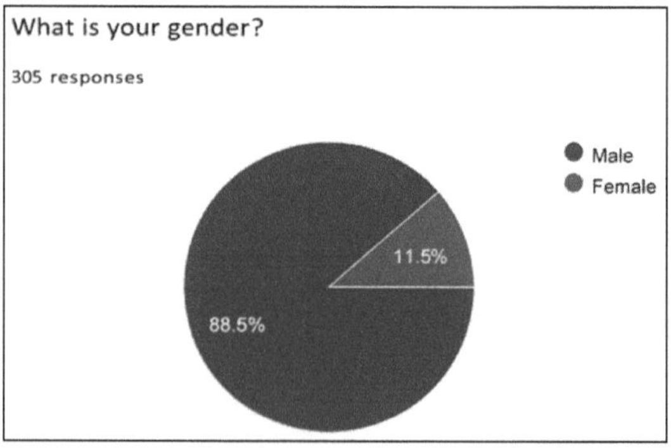

FIGURE 11.5 The gender of the responders.

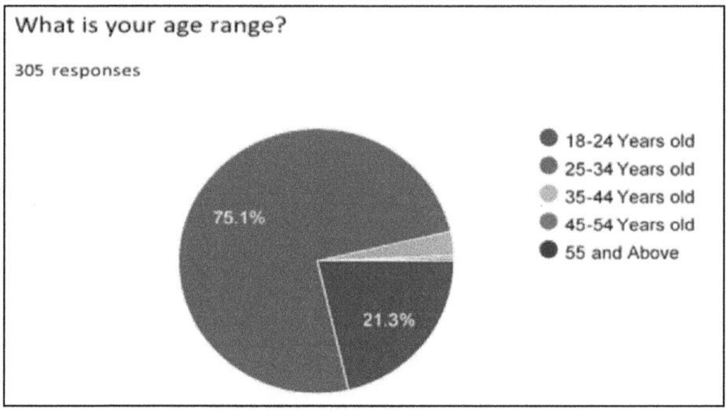

FIGURE 11.6 The age of the responders.

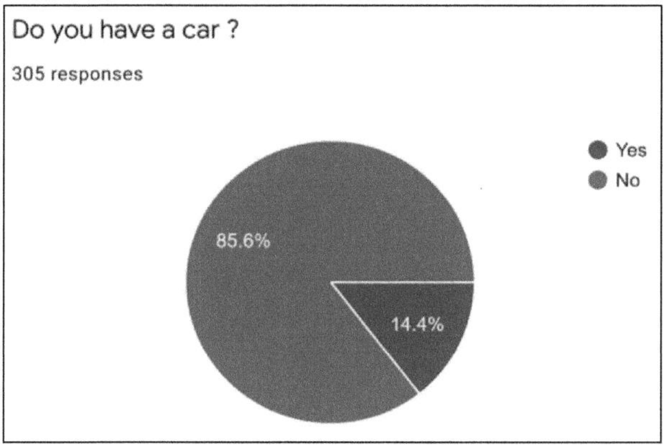

FIGURE 11.7 The possession of private vehicles by responders.

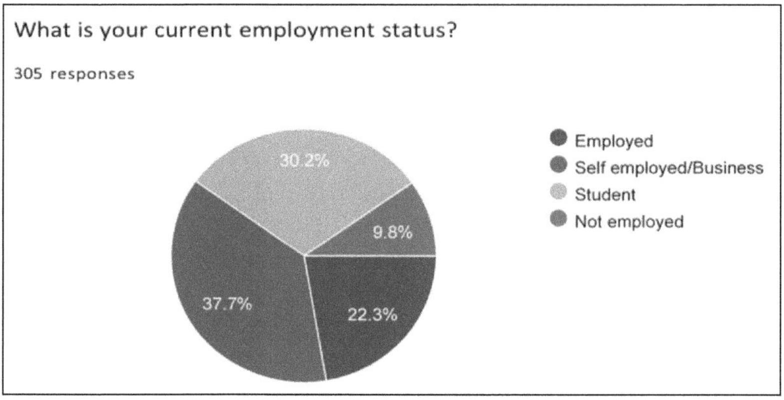

FIGURE 11.8 The employment status of the responders.

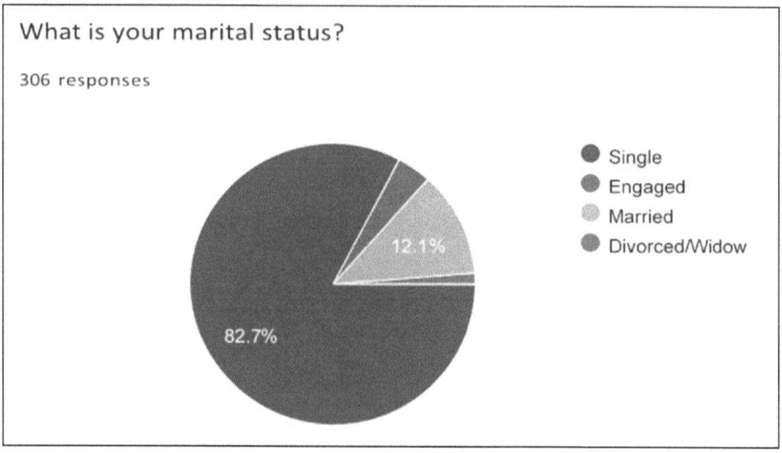

FIGURE 11.9 The marital status of the responders.

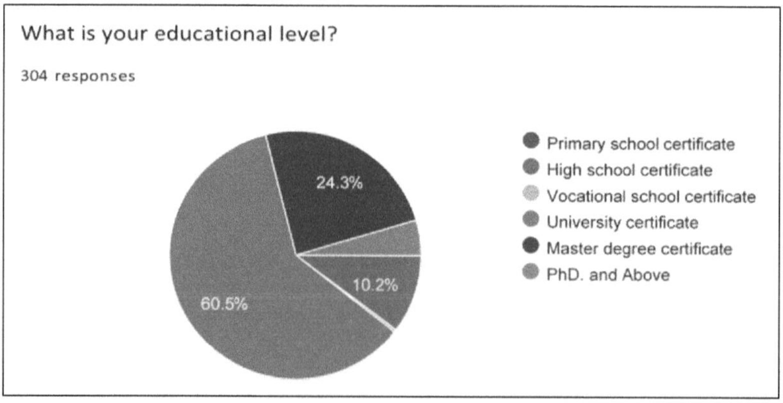

FIGURE 11.10 The educational level of the responders.

11.10.3 SECOND SECTION: SUITABILITY OF ADAPTING ARTIFICIAL INTELLIGENCE (AI)-BASED GREEN TRANSPORTATION

From the data recorded and analysis conducted, it can be concluded that most of the existing transportation modes are being operated with the use of fossil fuels such as gas, oil, coal, etc., to power the engine for operation. But when considering AI -based green transportation systems, it is among the key factor for having a sustainable green environment with minimum damages from human daily activities. This smart transportation system

(STS) sensitizes the environment from any future detriment that might likely arise due to human activities and makes it friendly with nature. A mere example of the significance of using the green transportation system can be seen from the respondent's opinions in using these green forms of transportation. For example, using an electric car that requires no burning of fuel for its operation will save the environment from the excess emission of the car exhaust to the atmosphere which will reduce the level of pollution of our environment (Figures 11.11–11.20).

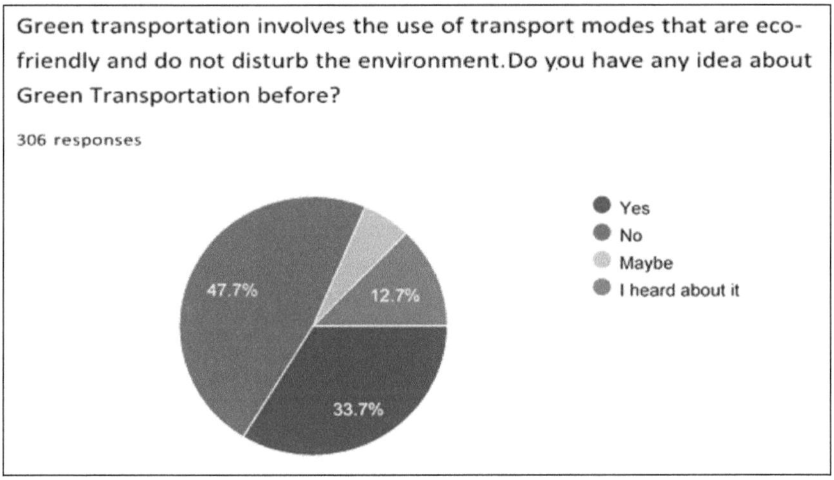

FIGURE 11.11 The knowledge of the responders on green transportation.

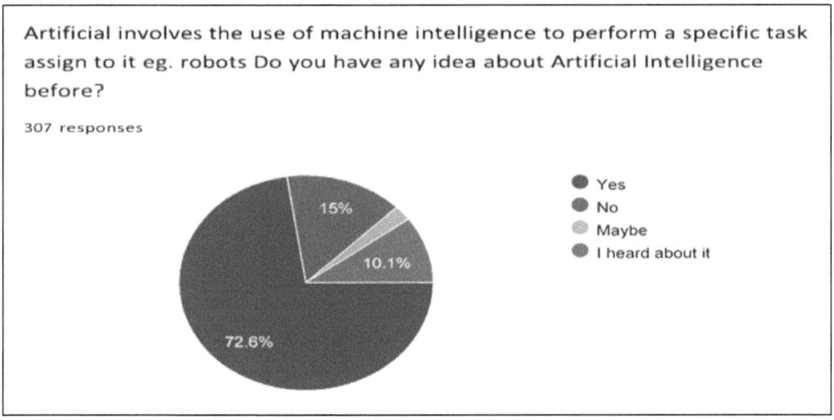

FIGURE 11.12 The knowledge of the responders on artificial intelligence.

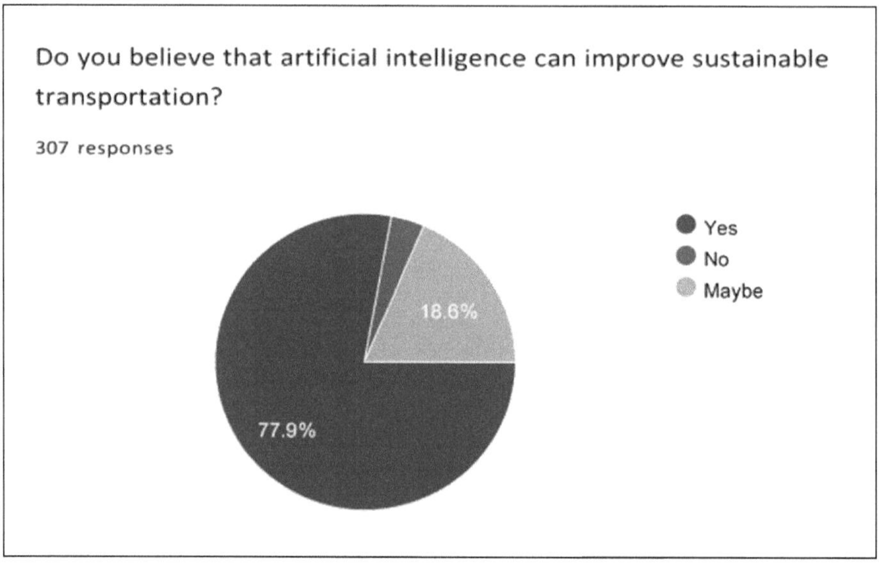

FIGURE 11.13 The knowledge of the responders about the application of artificial intelligence in green transportation.

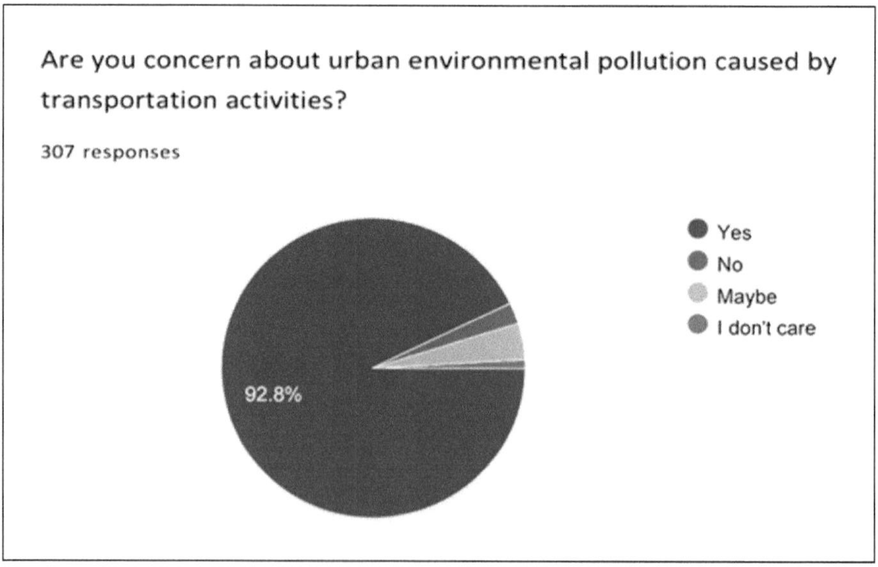

FIGURE 11.14 The knowledge of the responders about the pollution mitigation on urban pollution.

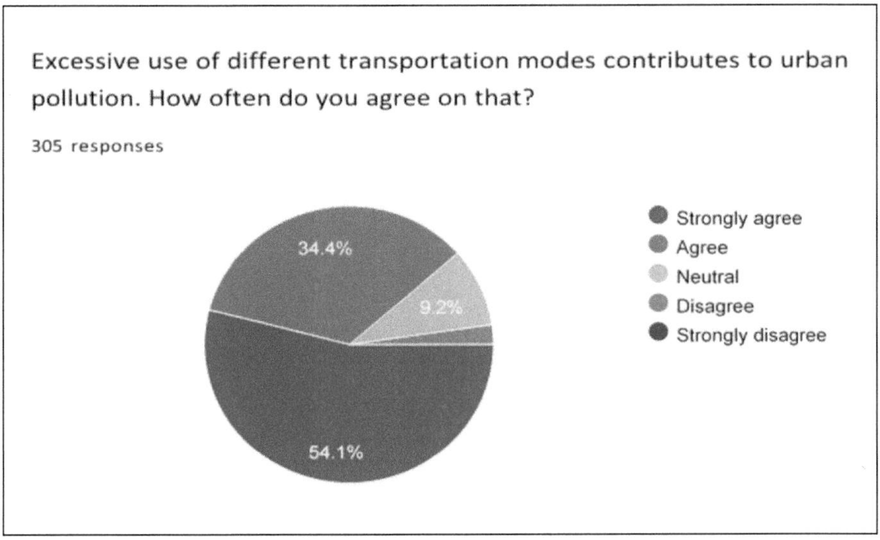

FIGURE 11.15 The responder's perception on pollution caused by conventional transportation modes.

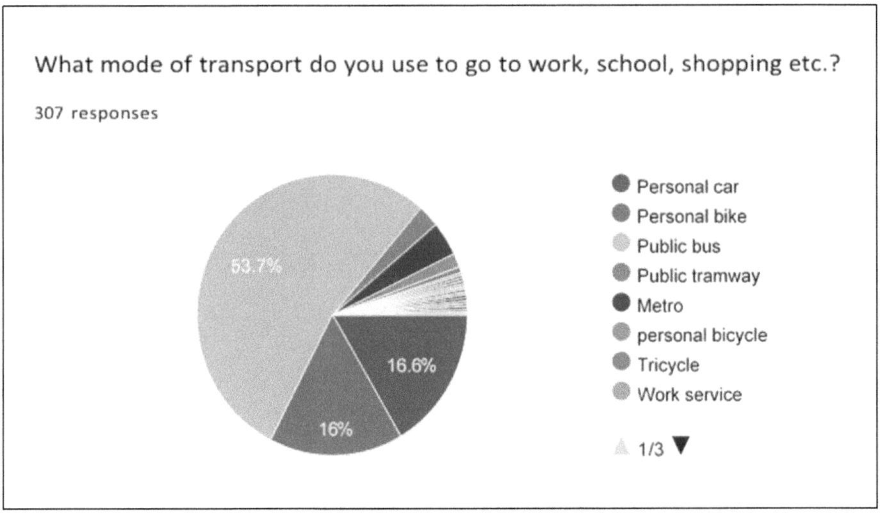

FIGURE 11.16 The usual modes of transportation responders.

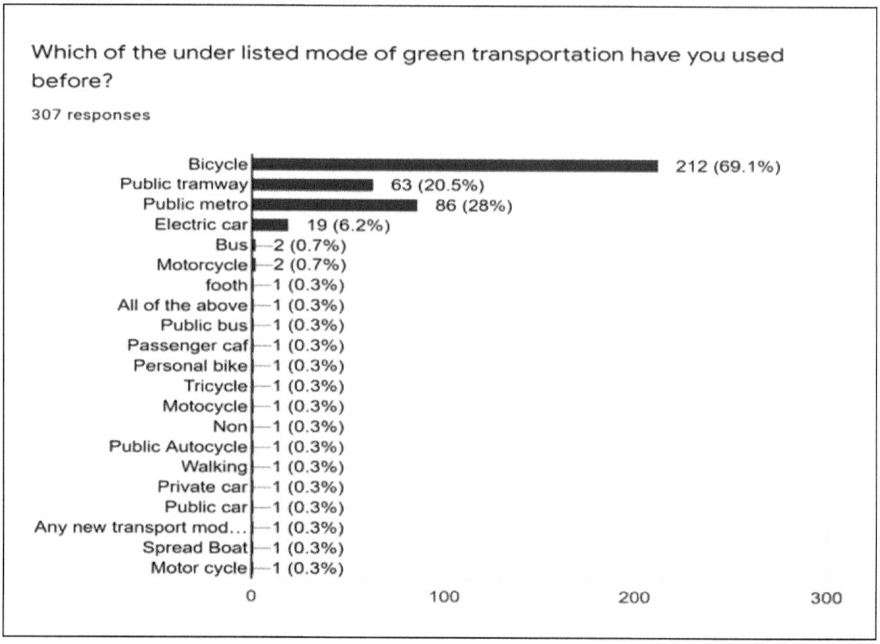

FIGURE 11.17 Green transportation systems utilized by the responders.

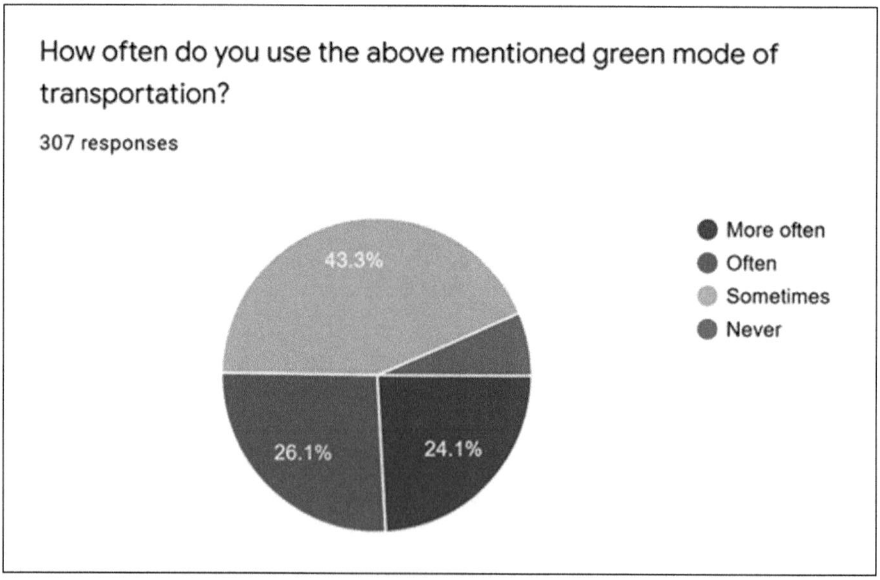

FIGURE 11.18 The extent of using the green transportation by the responders.

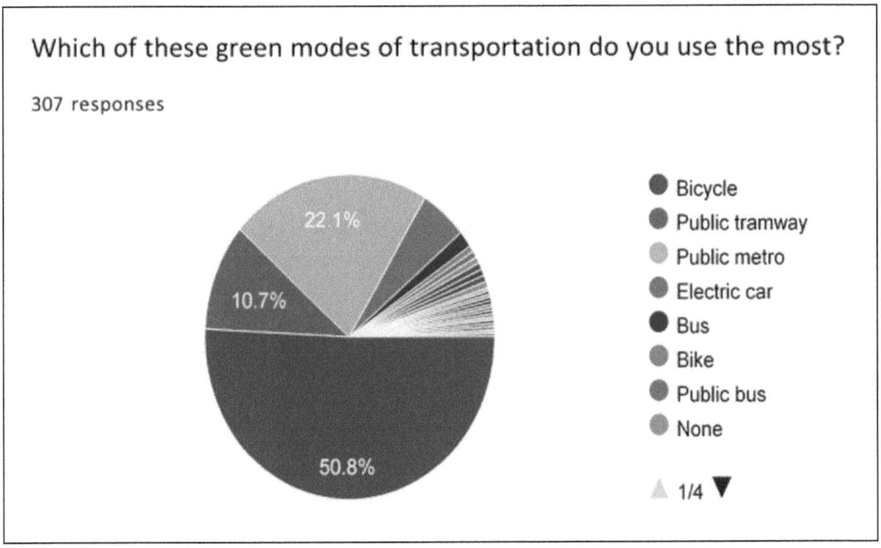

FIGURE 11.19 The extent of using the green transportation vehicles by the responders.

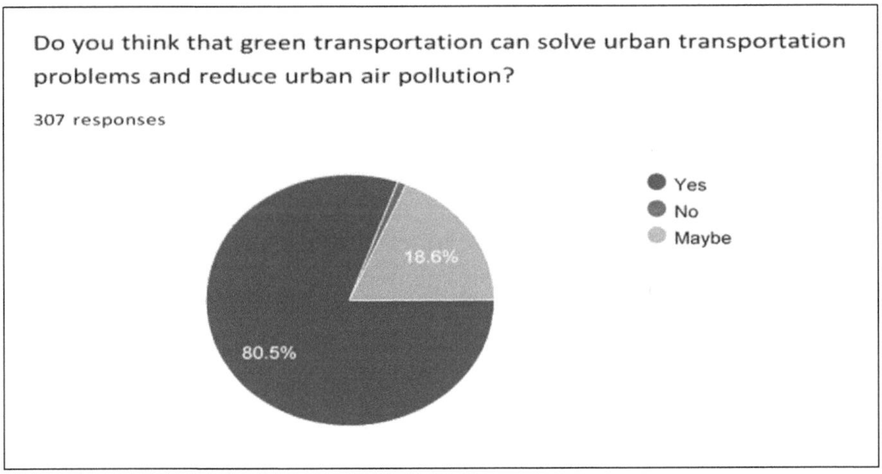

FIGURE 11.20 The perception of the responders to the implementation of green transportation.

The use of bicycles by people in the urban areas instead of driving personal cars to work, school, or marketplaces will both increase the health condition of the people as well as the environment. Using green trains in place of coal trains, walk in a walkable distance to school, grocery stores or markets will also improve the health of people and their environment. In general terms, the use of these green transportation systems has a great impact on the sustainable environment by making it a pollution-free urban environment, improve the health condition of people, boost economic and individual resources by using the less money defendant green facilities among others.

11.11 CONCLUSION

People living in the urban areas shall aim at developing sustainable transport systems. This will enable them to achieve the appropriate level of growth both in economic, social, and environmental spheres. Because, green transportation advocates the decreased use of private motor cars, the increased use of walking, bicycle, and public transport, and the use of clean energies and vehicles. It is also a low-cost, pollution-free, land resource and space saving transportation system suitable for all kinds of travelers.

Lack of proper implementation is the most challenging task toward achieving the goals of sustainable transportation, these challenges were hindered by mere regional, cultural, and in most cases by political reasons. Therefore, better understanding of these factors causing the setback is much important, these can be achieved through public consultation of the concerned communities, regulating the current traffic policies, pricings, and managements to suite the targeted sustainability requirements as well as the consideration of cultural setup of the concerned communities. These proposed measures toward implementation will probably take time and need resources, which in turn will avoid unreasonable delay leading to the collapse of the system in general. Political support perhaps is crucial toward the implementation of sustainable transportation ideas to the optimal required level, while effective and consistent enforcement is an essential part of implementation. Although, sometimes enforcement found to be problematic, where enforcement proves problematic, modification of the regulation, charge, etc., can be used to render enforcement [66].

KEYWORDS

- **bus rapid transit**
- **elevated light rail transit**
- **global fuel economy initiatives**
- **green transportation**
- **light rail transit**
- **sustainability**

REFERENCES

1. Loannis, M., Elisavet, S., Agathangelos, S., & Eugenia, B., (2020). Environmental and health impacts of air pollution: A review. *Frontiers in Public Health, 8*(14). doi: https://doi.org/10.3389/fubh.2020.00014.
2. Federico, K., Claudio, A. B., Carlos, F. C. D., Annette, M. P.-U., Sophie, B., Adair-Rohani, H., & Markus, A., (2015). Contributions to cities' ambient particulate matter (PM): A systematic review of local source contributions at global level. *Atmospheric Environment, 120,* 475–483. https://doi.org/10.1016/j.atmosenv.2015.08.087.
3. *World Urbanization Prospect*, (2014). The 2014 Revision. Department of Economic and Social Affairs, United Nations. New York.
4. Jeffrey, D. S., (2015). The age of sustainable development. *Journal of American Planning Association, 81*(3), 241, 242. https://doi.org/10.1080/01944363.2015.1077080.
5. *World Population Prospects*, (2010). The 2010 Revision. United Nations Department of Economic and Social Affairs (UNDESA) New York.
6. World Energy Scenarios, (2007). *Transport Technologies and Policy Scenarios*. World Energy Council. United Kingdom. http://www.worldenergy.org/publications/809.asp (accessed on 1 January 2023).
7. Luell, B., (2004). *Wildlife and Traffic–A European Handbook for Identifying Conflicts and Designing Solutions*. Transportation Research Board.
8. Lloyd, W., (2012). *Transport for Green Cities: An Urban Development Series* (pp. 175–217). Asian Development Bank.
9. Schipper, L., Fabian, H., & Leather, J., (2009). Transport and carbon dioxide emissions: Forecasts, options analysis, and evaluation. *Sustainable Development, 9*, 1–57.
10. Roberts, L. G., & Patterson, T. J., (2014). *Biofuels Encyclopedia of Toxicology* (3rd edn., pp. 469–475). doi: https://doi.org/10.1016/jB978-0-12-386454-3.01054-X.

11. Zuidgeest, M. H. P., Witbreuk, M. J. G., & Van, M. M. F. A. M., (2000). Sustainable transport: A review from a pragmatic perspective. *South African Transport Conference 'Action in Transport for the New Millennium' Conference*. South Africa.
12. Our Common Future, (1987). *World Commission on Environment and Development.* Oxford University Press, Oxford.
13. Leal, W., (2000). Dealing with misconception on the concept of sustainability. *International Journal of Sustainability in Higher Education, 1*(1), 9–19.
14. Dragun, A. K., & Jakobsson, K. M., (1997). *Sustainability and Global Environmental Policy: New Perspectives*. Edward Elgar, Cheltenham.
15. *Innovating for Sustainable Growth: A Bioeconomy for Europe*, (2012). European Commission. Brussels.
16. Sajida, P., Md. Kamruzzaman, & Tan, Y., (2017). Developing policy scenarios for sustainable urban growth management: A Delphi approach. *Sustainability, 9*(10), 1787. https://doi.org/10.33390/su9101787.
17. Ogryzek, M., Adamska-Kmieć, D., & Klimach, A., (2020). Sustainable transport: An efficient transportation network—Case study. *Sustainability, 12*(19), 8274. https://doi.org/10.3390/su12198274.
18. Han-Ru, L., (2016). Study on green transportation system of international metropolises. *Procedia Engineering, 137*, 762–771.
19. Yin, G. T., & Li, Q., (2009). Planning and practice of green transportation system – taking Sino-Singapore Tianjin eco-city as an example. *City Transportation, 4*, 56–58.
20. Penalosa, E., (2003). *Reader on World Transport Policy and Practice*. London: Earthscan.
21. Green, Hong Kong, (2011). *Environmental Bureau, Environmental Protection Department.*
22. Banister, D., (2008). The sustainable mobility paradigm. *Transp. Policy, 15*, 73–80.
23. Stringer, L. C., Fraser, E. D. G., Harris, D., Lyon, C., Pereira, L., Ward, C. F. M., & Simelton, E., (2020). Adaptation and development pathways for different of farmers. *Environmental Science and Policy, 104*, 174–189. https://doi.org/10.1016/j.envsci.2019,10.007.
24. Klinger, T., Kenworthy, J., & Lanzendorf, M., (2013). "Dimensions of urban mobility cultures–a comparison of German cities. *Journal of Transport Geography, 31*, 18–29. doi: https://doi.org/10.1016/j.jtrangeo.2013.05.002.
25. Santos, G., Behrendt, H., & Teytelboym, A., (2010). Part II: Policies instruments for sustainable road transport. *Research in Transportation Economics, 28*(1), 46–91.
26. Mu, R., & Jong, M. De., (2012). Establishing the conditions for effective transit-oriented development in China: The case study of Dalian. *Journal of Transport, Geography, 24C*, 234–349. doi: https://doi.org/10.1016/j.jtrango.2021.02.010.
27. United States Department of State (2010). *US Climate Action Report 2010*. Available at: http://unfccc.int/resource/docs/natc/usa_nc5 (accessed on 1 January 2023).
28. Luis, V., Nora, E. M., Andrea, G. Z., Sara, P., Bernd, D., & Biagio, G., (2015). Sustainable transportation strategies for decoupling road vehicle transport and carbon dioxide emissions. *Management of Environmental Quality: An International Journal, 26*(3), 373–388. doi: https://doi.org/10.1108/MEQ-07-2014-012.
29. Heinrichs, H., Jochem, P., & Fichtner, W., (2014). Including road transport in the EU ETS (European Emissions Trading System): A model-based analysis of the German

electricity and transport sector. *Energy, 69*, 708–720. doi: https://doi.org/10.1016/j.energy.2014.03.061.
30. Zuidgeest, M. H. P., & Van, M. M. F. A. M., (2000). Transportation planning for sustainable development. *Proceedings South-African Transportation Conference 2000*, Pretoria, South-Africa.
31. *The Future We Want*, (2012). Rio de Janeiro: United Nations.
32. *Global Status Report on Road Safety*, (2015). WHO Library Cataloguing-in-Publication Data, 2015.
33. Jeani-Paul, R., Claude, C., & Brain, S., (2006). *The Geography of Transportation System*. Taylors and Francis group.
34. Dockery, D., & Pope, C., (1994). Acute respiratory effects of particulate air pollution. *Annual Review Public Health, 15*, 107–132.
35. Jos, G. J., Olivier, G., Jeroen, J. M., & Peters, A. H. W., (2012). *Trends in Global CO_2 Emissions*. PBL Netherlands Environmental Assessment Agency, The Hague.
36. Evans, G., (2009). Creative cities, creative spaces and urban policy. *Urban Studies, 46*(5, 6), 1003–1040. doi: https://10.1177/0042098009103853.
37. World Health Organization (WHO), (2004). *World Report on Road Traffic Injury Prevention: Summary.* Geneva.
38. World Report on Road Traffic Injury Prevention: Summary, (2008). *The Global Burden of Disease: 2004*. World Health Organization (WHO). Geneva.
39. IEA Mobility Model, (2006). *Unpublished IEA Database*. Paris: IEA. International Energy Agency (IEA). Statistics and Balances. Paris: IEA.
40. Clarke, L., & Calvin, K., (2008). *Sectoral Approaches in International and National Policy*. Presentation for the 2nd International Expert Meeting on Bottom-up Based Analysis on Mitigation Potential.
41. Schrank, D., Lomas, T., & Eisele, B., (2011). *2011 Urban Mobility Report*. College Station: Texas Transportation Institute.
42. Willoughby, C., (2000). *Managing Motorization*. World Bank Discussion Paper. TWU-42. Washington, DC: World Bank.
43. Qiu, B., (2007). Fulfilling bus transportation priority strategies, expediting the construction of socialist harmony. *Urban Vehicles*, (1), 27–33.
44. Wright, L., (2005). *Car-Free Development*. Eschborn: GIZ.
45. Du Plessis, C., (2002). *Agenda 21 for: Sustainable Construction in Developing Countries–A Discussion Document.* Available at: http://researchspace.csir.co.za/dspace/handle/10204/3511 (accessed on 1 January 2023).
46. Ehsani, M., Gao, Y., & Emadi, A., (2010). *Modern Electric, Hybrid Electric and Fuel Cell Vehicles-Fundamentals, Theory and Design* (2nd edn., pp. 1–18). CRC Press, New York.
47. Rebecca, M., (2014). *The History of the Electric Car*. U.S. Department of Energy.
48. https://www.peoplespostmedia.com (accessed on 1 January 2023).
49. https://www.irena.org/-/media/files/IRENA/Agency/Publication/2016/IRENA_REmap_2016_edition_report.ashx (accessed on 4 February 2023).
50. Jiang, Y., & Li, Z., (2010). *Practices and Policies of Green Urban Transport in China.* Journeys.
51. Alberto, B., (2003). *Traffic Congestion the Problem and How to Deal With.* United Nation Publication.

52. Ian, S., (2016). The policy of sustainability and development. *Annual Review of Environment and Resources, 41*, 293–319. doi: https://doi.org.10.1146/annurev-environ-110625-090039.
53. Tsigdinos, S., Parsskevopoulos, Y., & Rallatou, N., (2019). Transit oriented development (TOD) challenges and perspectives: The case of Athens metro line. In: *4th European Transport Conference*.
54. IEA, (2009). *Transport, Energy and CO_2: "Moving Towards Sustainability"*. International Energy Agency, Paris.
55. Annual Report, (2015). *United Nations Environmental Program (UNEP).*
56. Doust, K., & Black, J., (2018). *Visual Sustainability Metrics: A Community Discussion and Decision-Making Toll.* 2nd UITP Sustainable Development Conference. Milan. Italy.
57. Masoud, K., Hui, L., John, T. H., & Xiao, L., (2019). Application of permeable pavement in highways for storm water runoff management and pollution prevention: California research experiences. *International Journal of Transportation Science and Technology, 8*(4), 358–372.
58. Naumann, S., McKenna, D., Timo, K., Mav, P., & Matt, R., (2011). *Design, Implementation and Cost Elements of Green Infrastructure Projects.* Final report to the European Commission, DG Environment, Ecologic institute and GHK Consulting.
59. *After the Storm: How Green Infrastructure Can Effectively Manage Stormwater Runoff from Roads and Highways*, (2011). The Natural Resources Defense Council, NRDC, New York www.nrdc.org (accessed on 1 January 2023).
60. Trinomics, Alterra, Arcadis, (2014). *Supporting the Implementation of the European Green Infrastructure Strategy*. EC DG ENV, ENV.B.2/SER/2014/0012. Regional Environment Centre, Risk & Policy Analysis, Stella Consulting.
61. McKinnon, A., (2012). "The role of government in promoting green logistics." In *Green Logistics: Improving the Environmental Sustainability of Logistics*, 372–396, 3rd Edition, Kogan page Limited.
62. Čepinskis, J., Aplinkos, T., & Masteika, I., (2011). Impacts of globalization on green logistics Centre's in Lithuania. *Environment Research, Engineering and Management, 1*(55), 34–42.
63. Abhishek, K., Manju, P., Pooja, D., & Jyotir, M. C., (2020). Framework for realization of green smart cities through the internet of things (IoT). *Trends in Cloud-based IoT: Innovations in Communication and Computing, 6*, 86–101.s https://doi.org/10.1007/978-3-030-40037-8.
64. Chourabi, H., Nam, T., Walker, S., Gil-Garcia, J. R., Mellouli, S., Nahon, K., & Scholl, H. J., (2012). Understanding smart cities: An integrative framework. In: *45th Hawaii International Conference on System Sciences* (pp. 2289–2297).
65. Irina, M., Rifat, K., Ksenia, S., & Aleksey, B., (2016). *Ensuring Sustainability of the City Transportation System: Problems and Solutions.* ICSC ES Web of Conferences, 02004. doi: https://10.1051/e3sconf/20160602004.
66. Sustainable Transport Policies, (1995). *European Conference of Ministers of Transport (ECMT).* Urban travel and sustainable development, ECMT, Paris. https://www.itf-oecd.org/sites/default/files/docs/00sustain_0.pdf (accessed on 1 January 2023).

Index

B

C

D

E

M

Q

R

U

V

W

X

Z